INTRODUCTORY STATISTICS

INTRODUCTORY STATISTICS

M. H. QUENOUILLE, Sc.D.
Imperial College of Science and Technology,
London

THE QUEEN'S AWARD TO INDUSTRY 1966

PERGAMON PRESS

OXFORD · LONDON · EDINBURGH · NEW YORK
TORONTO · SYDNEY · PARIS · BRAUNSCHWEIG

Pergamon Press Ltd., Headington Hill Hall, Oxford
4 & 5 Fitzroy Square, London W.1
Pergamon Press (Scotland) Ltd., 2 & 3 Teviot Place, Edinburgh 1
Pergamon Press Inc., Maxwell House, Fairview Park, Elmsford, New York 10523
Pergamon of Canada Ltd., 207 Queen's Quay West, Toronto 1
Pergamon Press (Aust.) Pty. Ltd., 19a Boundary Street,
Rushcutters Bay, N.S.W. 2011, Australia
Pergamon Press S.A.R.L., 24 rue des Écoles, Paris 5e
Vieweg & Sohn GmbH, Burgplatz 1, Braunschweig

First edition 1950
Reprinted 1966
Reprinted 1969

Library of Congress Catalog Card No. 65-28318

PRINTED IN GREAT BRITAIN BY A. WHEATON & CO., EXETER

08 009000 1

CONTENTS

CONTENTS

CONTENTS

PREFACE

THIS book has been written to provide a connected account of the more common statistical tests. Its writing was prompted by the belief that an elementary non-mathematical manual on statistics could serve the two purposes of supplying students of statistics with a textbook and of helping research workers to a fuller understanding of the methods used in the analysis of their results. The opinion that such a book was needed has often been expressed to me by students and research workers and in attempting to satisfy this need I have made use of parts of my lectures and, wherever possible, actual problems and data encountered in research work.

Statistical theory is largely mathematical, but no wide mathematical knowledge is required to understand the tests and methods derived from this theory. Unfortunately, the mathematical basis of statistical theory often discourages any attempts at understanding the assumptions behind statistical methods and appreciating statistical concepts. Some parts of the subject are undoubtedly mathematical, but most may be appreciated with a relatively small knowledge of mathematics. It is for this reason that the chapters in this book have been divided into two parts, the first part giving elementary applications of statistical methods and the line of reasoning involved in their use while the second part gives elementary parts of statistical theory and more advanced applications. It is hoped that by this means the interested student will be able to gain a connected account of the subject and to appreciate the bases of statistical methods by reading the first section of each chapter. Paragraphs numbered A contain the more specialized and advanced applications and may be included or omitted without breaking the continuity of the text.

Thanks are due to Miss M. J. ROBERTSON and Miss C. CARDNO for their help in preparing the manuscript for press and to Mr B. B. PARRISH for assistance with the proof reading. I also wish to thank the publishers and printers for the care they have taken in producing this book.

M. H. QUENOUILLE

Department of Statistics
University of Aberdeen
April 1950

INTRODUCTION

It is fitting to start a book of this kind with an explanation of the purpose of statistical science. Statistics is normally defined as the collection, presentation, and interpretation of data. These three purposes are interwoven and controlled one by another to a large extent, but until the beginning of this century far more attention was paid to the collection and presentation of data than to their interpretation. Masses of statistics were often collected and frequently misinterpreted if indeed interpretation was attempted. However, since that time the importance of a scientific approach in the interpretation of data has been realized and great steps have been made in the development of appropriate methods. Correspondingly, methods of collection and presentation of data have been altered to keep pace with the new methods of interpretation until, at present, the interpretation of data holds a central position in statistics, and methods of collection and presentation are hinged upon methods of interpretation. Hence, in this book, emphasis will be laid upon new methods of interpretation and their consequences, rather than on long standing methods of collection and presentation.

The growth in the methods of interpretation of data during the past fifty years may be linked with the names of W. S. Gosset, K. Pearson and, in particular, Sir Ronald Fisher. This growth started in the field of biological research, but the nature of the problems encountered has caused the new methods to be applied to medical, psychological and economic data, and to a lesser extent in physics and engineering. The biologist in his routine work is confronted with the difficulty that the measurements after identical treatment of two animals or plants, apparently similar in all respects, can give widely differing results. For example, an insecticide applied to two batches of insects can give different percentage kills, while the change in crop yield due to the application of a fertilizer may vary widely from field to field. This natural variability more or less masks any real effect, and makes the drawing of conclusions from collected data difficult, so that while the physicist can frequently use a single observation as a basis for further work the biologist can seldom rely upon one observation. Thus it was in the biological field that the laws of variability and their applications to the interpretation of data were initially investigated, but the general nature of these laws has made the new statistical techniques of much wider use.

In this text examples have been chosen from most fields of statistical application. However, since biological data are free from many of the complicating considerations which often occur in economic and psychological data, a high proportion of biological examples has been used. Nevertheless, these examples may be taken as demonstrating methods basically the same in application to all sciences.

It would be unfortunate if the emphasis laid upon methods of interpretation in these examples led the reader to the erroneous conclusion that a statistical test was always required in the interpretation of data or that only results which the statistician regarded as significant were important. While application of a statistical test helps to decide the accuracy of the data and the reality of effects, visual inspection is often all that is required to reach reliable results. This does not make the statistical test less useful, but it should be regarded as a tool to assist in the interpretation of data and not as the ultimate goal of every numerical investigation.

I

PRESENTATION OF SETS OF MEASUREMENTS

1.1 *Sampling*—If a series of measurements is made, it is implied that there is a variability in the results which necessitates taking more than one measurement. This variability may be due to the measurements being taken on different objects or under different conditions, as with a set of children's weights, or due to errors or changes in the method of measurement, as might occur in the evaluation of a physical constant such as g, or due to a combination of these, as with plot yields in an agricultural lay-out. If a set of children's weights is taken, this is commonly described as a 'sample' from a particular population, unless every child in the population is measured. This terminology is extended in statistical work so that a set of measurements of the physical constant g is said to be a sample from the population of possible measurements of g, and plot yields are said to be samples from the population of plot yields grown on similar soils under similar conditions. In this wider sense the population does not necessarily exist. For example, it is impossible to measure the yield of a field-plot more than once under the same conditions, since measurements will be affected by the drain on the resources of the land caused by previous measurements. It is nevertheless useful to think of a plot yield as a sample from the population of possible plot yields.

1.2 *Ordering the sample*—A sample is usually taken from a population in order to derive some conclusion concerning the population, and we must therefore search for a method of presentation which will conveniently summarize the properties of the population.

First, the sample might be arranged in order of magnitude so that, for example, if the weights of children have been taken in pounds, this arrangement might give 82, 87, 89, 91, 92, 92, 93, 93, 93, 94, 94 . . . This, while preserving all the information in the sample, presents it in a form more suitable for any subsequent work *e.g.* the proportion of children below any particular weight can readily be calculated. This presentation is very lengthy, especially if the sample is large. Although it is possible to shorten this to some extent by indicating in parentheses after each weight the number of children having that weight and omitting the parentheses when only one child is observed with a particular weight *e.g.* 82, 87, 89, 91, 92(2), 93(3), 94(2) . . . the length of this method of presentation will usually make some concise form of summary necessary.

1.3 *Grouping*—The children's weights in the above example will have been taken to the nearest pound or possibly to the nearest ounce, it being assumed that differences of less than a pound or an ounce will not be important. This gives a possible method of shortening the presentation of data, namely by grouping. If the weights are given to the nearest 5 lb, then the presentation is more concise than previously although some information has been lost. For example, the children's weights might now be 85(2), 90(4), 95(12), 100(20) . . . but it is now impossible to distinguish between children of weight 97 lb 9 oz and 102 lb 7 oz since they will both be given weights of 100 lb in the presentation. While this might not matter, it is obvious that an attempt to abbreviate the data still further by grouping, say, to the nearest 20 lb might result in the loss of important information. It is therefore necessary to compromise between the increase in brevity and the decrease in accuracy in choosing the width of the grouping or, as it is more commonly called, the grouping interval.

The manner in which a set of observations is distributed over the grouping intervals is called the 'frequency distribution' of the observations.

1.4 *Diagrammatic representation*—When the grouping interval has been decided, it is convenient to tabulate the data in a frequency table as shown in the first three columns of *Table 1.1*. In this table the grouping interval 82·5- contains all weights between 82·5 and 87·5 lb, including 82·5 lb but not 87·5 lb, and so on. Thus, strictly speaking, the mean weight in the group is only 85 lb if the weights can be determined exactly, but otherwise is slightly less than 85 lb. In this form the regularity or irregularity of any set of observations can be observed easily. The tabulation can also be shown in graphical form either by a histogram or by a frequency diagram, as in *Figure 1*. The histogram represents the frequency or number of measurements in each interval by a block, the area of which is proportional to the frequency, while the frequency diagram joins a set of points of which the distances from the axis are proportional to the various frequencies. It is seen that these two methods of representing data graphically are very similar, and that the narrower the grouping interval, the more similar are the outlines.

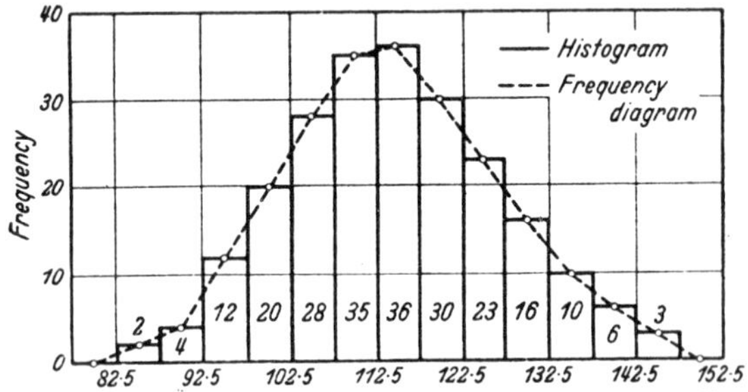

Figure 1. Histogram and frequency diagram of children's weights

The histogram is probably the more common

method of representation since, for example, the number of observations between any two values is represented by the area between those values in the histogram (the same is approximately true for the frequency diagram). For example, in *Figure 1* the number of observed weights between 100 and 120 lb is taken as $\frac{1}{2}\times 20+28+35+36+\frac{1}{2}\times 30=124$, as it is assumed that half the weights in the group centred on 100 lb are between 100 and 102·5 lb, and similarly for the group centred on 120 lb, half are assumed to have weights between 117·5 and 120 lb. In the same manner the number of observations between 88 and 103 lb is taken as $9/10\times 4+12+20+1/10\times 28=38$, this being the average obtained on the assumption that the observations in each group are spaced evenly throughout the interval; nine tenths of the observations in the group 87·5-92·5 lb being greater than 88 lb and one tenth of the observations in the group 102·5-107·5 lb being less than 103 lb.

Table 1.1. Frequency Table of Weights of 16-year old Children

Weight grouping lb	*Mean wt lb*	*Number of observations*	*Mean weight × number of observations*
82·5-	85	2	170
87·5-	90	4	360
92·5-	95	12	1,140
97·5-	100	20	2,000
102·5-	105	28	2,940
107·5-	110	35	3,850
112·5-	115	36	4,140
117·5-	120	30	3,600
122·5-	125	23	2,875
127·5-	130	16	2,080
132·5-	135	10	1,350
137·5-	140	6	840
142·5-	145	3	435
		225	25,780

Arithmetic mean = 114·58 lb

1.5 *The arithmetic mean and its calculation*—The frequency table is still a lengthy method of presentation and the comparison of sets of observations is not easy, so that a further method of condensation must be sought. One form of summary is obviously provided by the arithmetic mean or average of the set of measurements. Mathematically, it is convenient to denote the number of measurements by n, the values of the measurements by $x_1, x_2, x_3 \ldots x_n$ and the average measurement by $\bar{x}$. The average measurement is then given by the sum of the observations divided by the number of observations *i.e.*

$$\bar{x}=\frac{x_1+x_2+x_3+\ldots+x_n}{n}=\frac{\Sigma x}{n}$$

where Σx denotes the sum of the observations. This may be obtained directly from the observations, or alternatively if a great many observations are made, the frequency table can be used to shorten the calculation of the arithmetic mean with negligible loss in accuracy. If a frequency table is used the observations in each group may be taken at the centre of the group so that, for example, in *Table 1.1* the total of the 20 observations in the group centred on 100 lb is taken as 2,000 lb, the overall total as 25,780 lb and the mean as 25,780/225 = 114·58 lb.

Some of this calculation can be avoided by the use of short cut methods. If each observation is reduced by a constant amount, the mean will be reduced by this same amount, so that, if all observations are reduced by a quantity roughly equal to the mean, the calculations are less involved. For example, in *Table 1.2* the weights are reduced by 110 lb, and the multiplications are, as a result, shortened. The positive and negative halves of the table are added up separately, and an overall total of 1,030 lb is given at the bottom of the table. Thus the reduced mean weight is 1,030/225 = 4·58 lb, and the unreduced mean is 114·58 lb.

The calculation can be further reduced by using a new unit of measurement of 5 lb corresponding to the grouping interval. The method of calculation is given in *Table 1.3* in which the first and third columns are now one fifth of the corresponding columns in *Table 1.2*. The reduced mean is now 206/225 = 0·916 in units of 5 lb, so that the arithmetic mean is calculated as 110 + 5 × 0·916 = 114·58 lb as before.

Table 1.2. Use of Reduced Observations to Calculate the Arithmetic Mean

Mean weight −110	*Number of observations*	*Reduced mean weight × number of observations*
−25	2	−50
−20	4	−80
−15	12	−180
−10	20	−200
−5	28	−140
		−650
0	35	
5	36	180
10	30	300
15	23	345
20	16	320
25	10	250
30	6	180
35	3	105
	225	1,680
		−650
		1,030

Reduced mean = 1,030/225 lb = 4·58 lb

Unreduced mean = 110 + 4·58 lb = 114·58 lb

1.6 *Inadequacy of arithmetic mean*—Although the arithmetic mean is a concise method of presentation it is inadequate for several reasons; for example, it gives no indication of its reliability. A mean weight derived by measuring 1,000 children will usually be more reliable than a mean of two or three weights. However, not only the number of observations will affect the reliability of the arithmetic mean, but also the variability of the individual observations. For example, the mean weight of 10 children selected from a group with weights varying between 95 and 105 lb will be more reliable than the mean weight of 10 children selected from a group with weights varying between 80 and 120 lb. It is obvious that some index of the variability of the observations is required to overcome this difficulty.

Such an index would overcome in part other deficiencies of the arithmetic mean as a concise method of presentation. To say that the mean weight of a group of children is 114·58 lb gives no indication of the proportion of children above or below any particular weight since half of the children might weigh more than 115 lb (as is roughly true in the present example) or all of the children may have weights between 114 and 115 lb. This may be of particular importance in forestry, where it is frequently desired to find, say, the proportion of timber exceeding a certain size, or in industry, where the proportion of articles with quality or size falling below a fixed level may be required. Thus some measure of the scatter or spread of a set of measurements about their mean is required.

Table 1.3. Use of a Unit of 5 lb in Calculation of Arithmetic Mean

Adjusted wt =(*mean wt* −110)/5	*Number of observations*	*Adjusted wt* × *number of observations*
−5	2	−10
−4	4	−16
−3	12	−36
−2	20	−40
−1	28	−28
0	35	−130
1	36	36
2	30	60
3	23	69
4	16	64
5	10	50
6	6	36
7	3	21
	225	336
		−130
		206

Reduced mean =206/225 × 5 lb
=4·58 lb

Unreduced mean =110+4·58 lb
=114·58 lb

1.7 *Measures of spread: range and mean deviation*—Several measures can be used to indicate the spread or variability of a set of measurements. Of these the range, or difference between the highest and lowest observations, is the simplest and most easily calculated. For example, in *Table 1.1*, the range is 145 − 85 = 60 lb if the observations are assumed to fall at the centre of the grouping interval.

The range is not a satisfactory measure of the spread of a set of observations for two reasons. First, it cannot easily be used for comparative purposes since more extreme values will usually be observed as the number of observations is increased. The range will thus tend to increase as more observations are taken, but it is not clear how the range will be related to the number of observations. Secondly, the range can be distorted easily since it is determined by only two observations *e.g.* in *Table 1.1*, two extra measurements in the groups centred on 75 and 155 lb would hardly affect the set of observations but the range would be increased from 60 lb to 80 lb.

An alternative measure of spread is the mean deviation. This is calculated by taking the average difference between the mean and each of the observations irrespective of the sign of the differences. For instance, the set of ten observations 104, 107, 111, 112, 114, 115, 117, 119, 124, 127, has an arithmetic mean of 115, and the differences or deviations from the mean, irrespective of sign, are 11, 8, 4, 3, 1, 0, 2, 4, 9, 12, so that the mean deviation is

$$[11+8+4+3+1+0+2+4+9+12]/10=5{\cdot}4$$

Mathematically, the differences are denoted by $x_1-\bar{x}$, $x_2-\bar{x}$, $x_3-\bar{x}$ . . . and the disregard for the sign is expressed by parallels thus

$$|x_1-\bar{x}|,\quad |x_2-\bar{x}|,\quad |x_3-\bar{x}|,\ .\ .\ .$$

so that the mean deviation is

$$\frac{|x_1-\bar{x}|+|x_2-\bar{x}|+|x_3-\bar{x}|+\ .\ .\ .\ +|x_n-\bar{x}|}{n}=\frac{\Sigma|x-\bar{x}|}{n}$$

This measure of spread will not be affected by a few large or small observations in the same manner as the range. It can also be used for comparative purposes since it is virtually independent of the number of observations taken. (It is not completely independent of the number of observations, but provided the number of observations exceeds ten it is effectively so.) The magnitude of the mean deviation will indicate the region within which slightly more than half the observations will fall. In the above example slightly more than half the observations are within 5·4 of the mean 115 *i.e.* between 109·6 and 120·4.

To demonstrate the use of the range and mean deviation, suppose that 92, 99, 104, 110, 113, 115, 118, 121, 125 and 133 is a second set of ten observations. The arithmetic mean of this set of observations is 113 and the deviations from the mean, ignoring the signs, are 21, 14, 9, 3, 0, 2, 5, 8, 12 and 20, so that the mean deviation is

$$[21+14+9+3+0+2+5+8+12+20]/10=9{\cdot}4$$

Here, slightly more than half the observations lie between 113 − 9·4 and 113 + 9·4 *i.e.* 103·6 and 122·4. The spreads or scatters of the measurements in the two sets of observations are thus in the ratio 9·4/5·4 = 1·74, and scatter in the second set is 1·74 times that in the first. A comparison of the scatter in the two sets of observations using the range gives the ratio $\frac{133-92}{127-104}=\frac{41}{23}=1{\cdot}78$, which agrees remarkably well with this value.

1.8 *Measures of spread: standard deviation and variance*—The standard deviation is the most commonly used measure of spread and its square, which is called the variance, occurs almost as frequently. The variance is defined to be the average of the squared deviations of each observation from the arithmetic mean, or briefly, the mean squared deviation. Thus for the first set of observations given on p 6 the deviations are -11, -8, -4, -3, -1, 0, 2, 4, 9 and 12, and the variance is

$$[121+64+16+9+1+0+4+16+81+144]/10=45{\cdot}6$$

The standard deviation or root mean squared deviation is thus $\sqrt{(45{\cdot}6)}=6{\cdot}75$.

Under this definition the standard deviation and variance will not be much affected by a few large or small observations. For the purposes of comparison they will not be greatly influenced by the numbers of observations in each group although, if the number of observations tends to be small, the standard deviation and variance tend to be reduced*. This difficulty is overcome by altering the definition slightly, so that the variance is taken as *the total of the squared deviations divided by one less than the number of observations*†. Thus, in the above example, the variance by the revised definition is

$$[121+64+16+9+1+0+4+16+81+144]/9=50{\cdot}67$$

and the standard deviation is $\sqrt{(50{\cdot}67)}=7{\cdot}12$. It is impossible to justify completely the use of one less than the number of observations in calculating the mean squared deviation or variance without recourse to mathematics and, for this reason, a fuller mathematical justification is given on p 51. Since the deviations of the observations from the mean can be represented by $x_1-\bar{x}$, $x_2-\bar{x}$, $x_3-\bar{x}$, . . . $x_n-\bar{x}$, the definition of variance or mean squared deviation is

$$\frac{(x_1-\bar{x})^2+(x_2-\bar{x})^2+(x_3-\bar{x})^2+\ .\ .\ .+(x_n-\bar{x})^2}{n-1}=\frac{\Sigma\,(x-\bar{x})^2}{n-1}$$

and the standard deviation is $\left(\dfrac{\Sigma\,(x-\bar{x})^2}{n-1}\right)^{\frac{1}{2}}$.

If the second set of observations on p 6 is used, its variance is found to be

$$[441+196+81+9+0+4+25+64+144+400]/9=151{\cdot}56$$

and its standard deviation is $\sqrt{(151{\cdot}56)}=12{\cdot}31$. Thus, the ratio of the standard deviations of the two sets of observations is $12{\cdot}31/7{\cdot}12=1{\cdot}73$, which agrees with the values of 1·74 and 1·78 which were derived using the mean deviation and range.

It is usually true to say that about two thirds of any set of observations differ from the mean by less than the standard deviation. For the first set of ten observations roughly two thirds of the observations lie between $115-7{\cdot}12$ and $115+7{\cdot}12$ *i.e.* between 107·88 and 122·12, while for the

* For one observation, the standard deviation by this definition would be zero, whereas it is really impossible to get any information concerning scatter from only one observation.

† Under this definition, when only one observation is taken, the standard deviation is now 0/0, an indefinite quantity.

second set the corresponding limits are 100·69 and 125·31. As a result of this property, and because the standard deviation is obtained from a square root, it is prefixed by a plus or minus sign thus: $\pm 7{\cdot}12$, $\pm 12{\cdot}31$.

To assist with the calculation of the variance and standard deviation, *Table XI* in the Appendix gives the squares and square roots of numbers from 1 to 1,000.

1.9 *Sample estimates*—It was explained at the outset that a sample set of measurements is normally taken in order to draw conclusions about the population. In taking the arithmetic mean of a sample we are not finding, but estimating, the mean of the population; a second sample would, in general, have a different arithmetic mean. For example, if two groups of ten children of the same age are taken and the arithmetic means of their weights are found, then the two arithmetic means will usually be different, although they are measurements or estimates of the same quantity. Thus the mean of the population is called the true mean and is denoted by μ, while an arithmetic mean of a sample is called a sample mean or an estimate of the true mean and is denoted by $\bar{x}$. In general the more observations that are taken the more accurately will the true mean be estimated, while an increase in the population scatter will lead to less accurate estimation of the true mean. Thus the mean of a uniform population will be easier to determine accurately than the mean of a highly variable population.

When a sample set of observations has been selected it is usual to find the arithmetic mean in order to estimate the true mean of the population, but although the arithmetic mean of the sample gives the best estimate of the true mean it is not the only estimate. It is possible to take, for example, the average of the highest and lowest observations, or the central observation when the observations are arranged in order. In the sets of observations on pp 5, 6 the averages of the highest and lowest are 115·5 and 112·5, while the central observation would be taken as $(114+115)/2=114{\cdot}5$ and $(113+115)/2=114{\cdot}0$. Both of these methods might be used to estimate the true mean but, while the calculation is reduced, the estimate of the true mean is not as accurate as might be obtained by using the mean of the sample. In other words the arithmetic mean of a sample gives a more accurate value for the mean of the population than any other estimate (such as the mean of the highest and lowest observations) that might be used.

In the same manner, when the spread or scatter of the population is being estimated, it is possible to use several measures, such as range, mean deviation and standard deviation, and it can be proved mathematically that the standard deviation will usually give a more accurate estimate of the scatter in the population than any other measure of spread. For this reason, and for reasons which will be appreciated later, the standard deviation is commonly used as a measure of spread. Although the standard deviation involves more calculation than the range or mean deviation, the increase in accuracy warrants the extra calculation. Also it will be seen later that the calculation can be greatly reduced by the use of several short cut methods.

It is usual to denote the true value of the standard deviation in the population by σ, so that the true value of the variance is σ^2. The estimate of the standard deviation obtained from a sample is denoted by s, and the estimated variance by s^2. This practice of using Greek letters to denote properties of the population and the corresponding Roman letters to denote the estimates of these properties is commonly used in statistics.

A set of measurements can thus be summarized using the arithmetic mean and standard deviation but it is not at this stage obvious that this form of summary will suffice. It must be left until the next chapter to show that most of the properties of a sample can be found from its mean and standard deviation.

1.10 *Calculation of standard deviation and variance*—The calculation of the variance in the two sets of ten observations given above was simplified by the whole number values of the estimated means, but this will not normally occur and the calculation may be lengthened as a result. For example, the mean of the seven observations 104, 111, 112, 115, 117, 119, 124, is 114·57; the deviations, irrespective of sign, from this mean are 10·57, 3·57, 2·57, 0·43, 2·43, 4·43, 9·43 and the estimated variance is

$$\begin{aligned}&\tfrac{1}{6}[(10{\cdot}57)^2+(3{\cdot}57)^2+(2{\cdot}57)^2+(0{\cdot}43)^2+(2{\cdot}43)^2+(4{\cdot}43)^2+(9{\cdot}43)^2]\\&=\tfrac{1}{6}[111{\cdot}7249+12{\cdot}7449+6{\cdot}6049+0{\cdot}1849+5{\cdot}9049+19{\cdot}6249+88{\cdot}9249]\\&=\tfrac{1}{6}(245{\cdot}7143)=40{\cdot}9524\end{aligned}$$

It is seen here that the estimation of the standard deviation, which is $\sqrt{(40{\cdot}9524)}=6{\cdot}40$, is lengthened considerably as a result of the decimals introduced by the mean.

This difficulty can be overcome by a rule which may be stated as follows: the sum of the squares of the deviations of a set of observations from their mean is equal to the sum of the squares of the deviations from any convenient value less the squared sum of these latter deviations divided by the number of observations. This rule is proved on p 17. Mathematically, if a is any value, this rule is expressed by the equation

$$(x_1-\bar{x})^2+(x_2-\bar{x})^2+\ldots+(x_n-\bar{x})^2=(x_1-a)^2+(x_2-a)^2+\ldots+(x_n-a)^2-\frac{[x_1-a+x_2-a+\ldots+x_n-a]^2}{n}$$

or

$$\Sigma(x-\bar{x})^2=\Sigma(x-a)^2-\frac{[\Sigma(x-a)]^2}{n}$$

Table 1.4. Use of Reduced Observations to Calculate the Standard Deviation

Mean weight −110	Number of obser-vations	Reduced mean wt × number of observations	Reduced mean wt × third column
−25	2	−50	1,250
−20	4	−80	1,600
−15	12	−180	2,700
−10	20	−200	2,000
−5	28	−140	700
0	35	−650	0
5	36	180	900
10	30	300	3,000
15	23	345	5,175
20	16	320	6,400
25	10	250	6,250
30	6	180	5,400
35	3	105	3,675
	225	1,680	39,050
		−650	
		1,030	

$$\text{Estimated variance} = \frac{1}{224}\left[39{,}050 - \frac{1{,}030^2}{225}\right]$$

$$= \frac{34{,}335}{224} = 153{\cdot}28$$

Estimated standard deviation = 12·38 lb

where, as usual, $x_1, x_2 \ldots x_n$ denote the n observations, $\bar{x}$ denotes the estimated mean and the sign Σ indicates the summation of a series of terms. In particular, if a is chosen to be zero, we get

$$\Sigma (x-\bar{x})^2 = \Sigma x^2 - (\Sigma x)^2/n$$

If this rule is applied to the above example, a may be chosen to be 114, the deviations from this value are then −10, −3, −2, 1, 3, 5, 10, and the sum of the deviations, $\Sigma(x-a)$, is 4. The sum of squares of deviations from the mean, $\Sigma(x-\bar{x})^2$, is consequently equal to

$$10^2+3^2+2^2+1^2+3^2+5^2+10^2-\frac{4^2}{7} = 248-2{\cdot}2857$$

$$= 245{\cdot}7143$$

as before.

It should be noted that although the calculation is reduced when a is chosen near to the estimated mean it is not necessary for a to be so placed. For example, if a is chosen to be 110 the deviations become −6, 1, 2, 5, 7, 9, 14 with sum 32. The sum of squares of deviations from the mean is then

$$6^2+1^2+2^2+5^2+7^2+9^2+14^2-\frac{32^2}{7} = 392-\frac{1{,}024}{7}$$

$$= 392-146{\cdot}2857$$

$$= 245{\cdot}7143$$

This rule can also be used for grouped observations such as in *Table 1.1*. If a is taken here as 110, then there are two observations with a deviation

-25, four observations with a deviation -20, and so on, as in *Table 1.2*. The sum of squares of the deviations from 110 is

$$2\times(-25)^2+4\times(-20)^2+\ldots+3\times(35)^2$$
$$=(-50)\times(-25)+(-80)\times(-20)+\ldots+(105)\times(35)$$

and can be found by multiplying the first and third columns of *Table 1.2* together and adding. This has been done in *Table 1.4*. The variance is then estimated as

$$\frac{1}{224}\left[39{,}050-\frac{1{,}030^2}{225}\right]=153{\cdot}28$$

and the estimated standard deviation is 12·38 lb. This method can also be applied with a unit of 5 lb as in *Table 1.5* but it must be remembered that to adjust the variance to the normal unit of 1 lb it is necessary to multiply it by twenty five *i.e.* 5^2.

Since the term $[\Sigma(x-a)]^2/n$ corrects the sum of squares of deviations from an arbitrary value a to give the sum of squares of deviations from the estimated mean, it is often called the correction term.

Table 1.5. Use of a Unit of 5 lb in Calculation of the Standard Deviation

Adjusted wt =(mean wt − 110)/5	*Number of obser-vations*	*Adjusted wt × number of observations*	*Adjusted wt × third column*
−5	2	−10	50
−4	4	−16	64
−3	12	−36	108
−2	20	−40	80
−1	28	−28	28
0	35	−130	0
1	36	36	36
2	30	60	120
3	23	69	207
4	16	64	256
5	10	50	250
6	6	36	216
7	3	21	147
	225	336	1,562
		−130	
		206	

$$\text{Estimated variance}=\frac{1}{224}\left[1{,}562-\frac{206^2}{225}\right]$$
$$=\frac{1{,}373{\cdot}40}{224}=6{\cdot}1312$$

Estimated standard deviation = 2·476 in 5 lb units
= 12·38 lb

SUMMARY OF PP 1-11

The presentation of a set of measurements can be made by a frequency table or diagrammatically by a histogram or frequency diagram. These forms of presentation lack conciseness so that the arithmetic mean is used to indicate the average value of the measurements, while a measure of spread is used to indicate the variability of the measurements. There are three main measures of spread: range, mean deviation and standard deviation. Of these the latter is normally employed since it can be used to compare samples of different sizes and because it is usually a better estimate of the scatter of observations than any other measure. Thus a set of observations will be concisely presented by their mean and standard deviation.

EXAMPLES

1 The girths of 460 90 year-old Scots pines were tabulated with an 8 in grouping interval as follows:

Mean girth, in	25	33	41	49	57	65	73	81	
No. of trees	24	96	128	124	52	28	6	2	*Total*: 460

Construct a histogram and frequency diagram of this distribution, and show that the arithmetic mean and standard deviation are 44·51 and 10·78 in respectively.

A group of 24 European larches in the same area and of the same age had a mean girth of 39·50 in and a standard deviation of 9·75 in. Both the arithmetic mean and the spread of the girths of the European larches are less than those of the Scots pines but not appreciably so. In making comparisons such as these the possible inaccuracy of the arithmetic mean must be remembered; since the individual girths vary between wide limits it is likely that by chance this sample of 24 European larches included several trees of small girth. Thus any general conclusion drawn from these figures must take into account the accuracy of the arithmetic mean. The question of how to account for the variability of arithmetic means in reaching conclusions will be discussed in the next two chapters.

2 The scores of 100 male and 100 female students in a psychological test for assertiveness were tabulated with a grouping interval of ten as follows:

Mean score	−45	−35	−25	−15	−5	5	15	25	35	45	55	
Male frequency	1	6	8	15	22	15	18	8	4	1	2	*Total*: 100
Female frequency	1	4	7	4	12	19	21	12	10	6	4	*Total*: 100

Show that the mean score and standard deviation are 0·5 and 20·3 for the males and 10·7 and 22·5 for the females.

3 In ten batches of a hundred seeds, 11, 14, 10, 8, 10, 11, 4, 13, 5 and 14 seeds failed to germinate. Show that the means percentage failure is 10·0 and the standard deviation is 3·46.

In ten other batches of the same size from a different seed mixture, the numbers were 5, 7, 2, 6, 4, 5, 8, 6, 5 and 7. Here the mean and standard deviation are 5·5 and 1·72, so that this mixture gives a higher and less variable germination rate.

4 The frequency distribution of the number of species in each genus of *Acridiidæ* (short-horned grasshoppers) is given by C. B. Williams in *J. Ecol*, 32 (1944) 1, as:

No. of species per genus	1	2	3	4	5	6	7	8	9	10
Frequency	320	131	86	61	41	27	21	18	23	17
No. of species per genus	11	12	13	14	15	16	17	18	19	20
Frequency	12	8	9	3	5	4	3	6	2	3
No. of species per genus	21	22	23	24	25	26	27	28	29	30
Frequency	1	1	2	1	0	2	0	0	4	0

Demonstrate this distribution using a frequency diagram, and show that the mean and standard deviation of the number of species per genus are 3·925 and 4·542.

5 The following table gives indices of pig-iron production during four forty-month periods in each of which ten indices are calculated:

Sep 1915-Dec 1918	100·1	106·1	105·7	108·2	103·1	107·9	102·7	93·5	110·4	112·3
Jan 1919-Apr 1922	98·5	76·3	76·9	100·1	99·5	98·9	59·6	33·4	43·4	61·5
May 1922-Aug 1925	72·3	86·9	110·8	119·3	99·1	105·8	67·6	81·9	111·8	89·2
Sep 1925-Dec 1928	98·5	109·3	106·8	104·9	107·9	100·7	89·4	100·5	102·2	107·5

Verify that the arithmetic means and measures of spread are:

Period	*Mean*	*Range*	*Mean deviation*	*Standard deviation*
Sep 1915-Dec 1918	105·0	18·8	4·12	5·46
Jan 1919-Apr 1922	74·8	66·7	20·27	24·74
May 1922-Aug 1925	94·5	51·7	14·89	17·60
Sep 1925-Dec 1928	102·8	19·9	4·51	5·95

The means indicate clearly the drop in pig-iron production immediately after 1918 and the rise during the twenties. The measures of spread indicate that the drop was accompanied by a variability in the production of pig-iron which decreased as the production rose.

It should be noted again that the range is not a reliable measure of spread. In this example the range, mean deviation and standard deviation are related as follows:

Period	*Range ÷ standard deviation*	*Range ÷ mean deviation*	*Mean deviation ÷ standard deviation*
Sep 1915-Dec 1918	3·4	4·6	0·75
Jan 1919-Apr 1922	2·7	3·3	0·82
May 1922-Aug 1925	2·9	3·5	0·85
Sep 1925-Dec 1928	3·3	4·4	0·76

so that the ratio of range to mean deviation and standard deviation fluctuates widely, while the ratio of mean deviation to standard deviation is comparatively constant.

EXTENDED DEVELOPMENT

1A.11 *Other measures: median and mode*—It was suggested on p 8 that the central value of a set of observations might be used to estimate the mean. This value is called the median and it will provide an estimate of the mean if the observations fall symmetrically about the mean. If more or less than half of the observations are greater than the mean, the median will not serve as an estimate of the mean. The frequency table, the histogram, and frequency diagram are then said to be asymmetrical or skew, and the median provides a measure of what is usually called central tendency. *Figures 2* and *3* indicate skew histograms and frequency diagrams. To demonstrate the use of the median, consider the set of nine observations 80, 94, 110, 112, 113, 114, 116, 119, 123. The average of these observations is 109 but this cannot be called representative of the whole set of observations since it is exceeded by seven of the nine observations. The median, or fifth observation here, is 113 and this value is more indicative of the set of observations as a whole.

When the set of observations is asymmetrical, then the difference between the median and the mean indicates the extent of the deviation from symmetry, but since this difference varies with the spread of the observations or the scale on which they are measured it is usually divided by the standard deviation. Thus (mean − median)/(standard devn) is used as a measure of skewness. For example, the estimated variance of the above set of nine observations is $[29^2+15^2+1^2+3^2+4^2+5^2+7^2+10^2+14^2]/8=182{\cdot}75$ so that this measure has a value $(109-113)/(182{\cdot}75)^{\frac{1}{2}}=-0{\cdot}296$. The values of this index are also given in *Figures* 2 and 3, and it should be noted that the sign of this measure indicates the direction of the skewness.

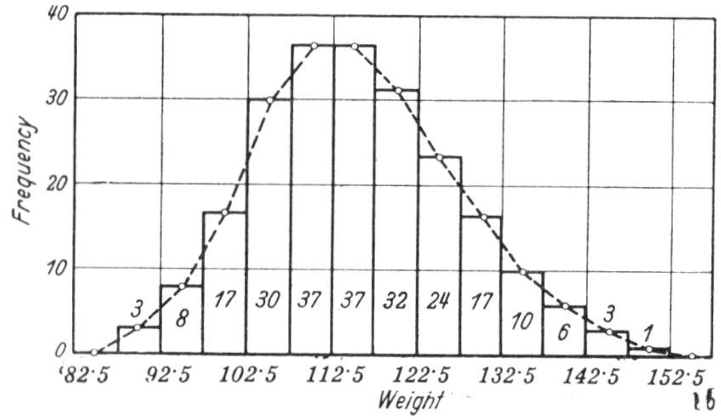

Mean = *115·69* *Standard deviation* = *11·84*
Median = *114·86* *Mode* = *112·50*
Quartiles = *107·42, 123·44*
Coefficient of variability = *10·23%*

$$\frac{\textit{Mean-median}}{\textit{Standard devn}}=0{\cdot}070$$

$$\frac{\textit{Mean-mode}}{\textit{Standard devn}}=0{\cdot}269$$

Quartile devn = *8·01*

Figure 2. Histogram and frequency diagram of children's weights

The median is easily calculated for a set of observations in a frequency table such as *Table 1.1*. Since there are 225 observations the 113th in order is required. 101 observations are less than 112·5 lb, so that the twelfth of the 36 observations in the group 112·5 − 117·5 is required. The value of this may be taken as $112{\cdot}5+11{\cdot}5\times5/36=114{\cdot}10$ lb, if it is assumed that the 36 observations are equally spaced

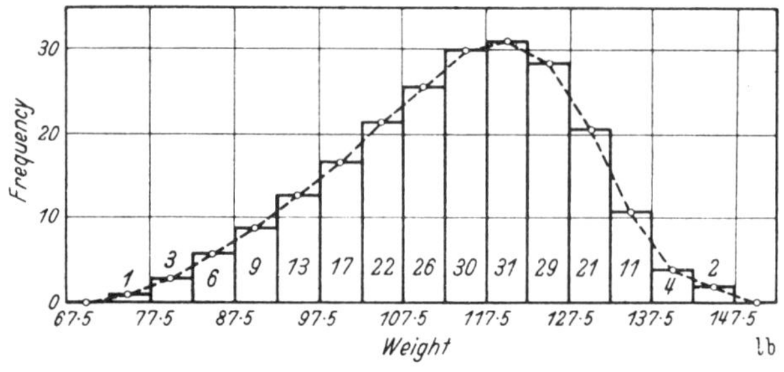

Mean = *113·78* *Standard deviation* = *14·16*
Median = *115·08* *Mode* = *119·17*
Quartiles = *104·20, 124·31*
Coefficient of variability = *12·45%*

$$\frac{\textit{Mean-median}}{\textit{Standard devn}}=-0{\cdot}092$$

$$\frac{\textit{Mean-mode}}{\textit{Standard devn}}=-0{\cdot}381$$

Quartile devn = *10·06*

Figure 3. Histogram and frequency diagram of children's weights

throughout the 5 lb interval. For this set of values the skewness is given by $(114{\cdot}58 - 114{\cdot}10)/12{\cdot}38 = 0{\cdot}039$ or roughly one half the skewness of *Figure 2* and one third the skewness of *Figure 3*.

Another measure of central tendency which is often employed when the set of observations is asymmetrical is the value corresponding to the most frequent observation. This can only be found for a frequency table, and is called the mode. For example, in *Table 1.1* more observations fall in the group 112·5 – 117·5 than in any other group, and the mode may be taken as approximately 115 lb. However, since more observations fall in the group 107·5 – 112·5 than in the group 117·5 – 122·5 the mode is in fact less than 115 lb*, but for most purposes the value of 115 lb is sufficiently accurate. For a symmetrical histogram or frequency diagram, the mean and mode will coincide (with the median), but if the set of observations is asymmetrical the difference between the mean and the mode may be used to indicate the extent of asymmetry. Thus (mean – mode)/(standard devn) is also used as an index of skewness, and this takes a value $(114{\cdot}58 - 113{\cdot}21)/12{\cdot}38 = 0{\cdot}111$ for the observations in *Table 1·1*. The values of this index are also given in *Figures 2* and *3*. In fact, the use of the median to indicate skewness is roughly equivalent to the use of the mode since in practice the latter index of skewness is usually about three times the former.

1A.12 *Further measures: quartiles and coefficient of variability*—In the previous section the median was defined as the value exceeded by one half of the observations. Correspondingly it is possible to introduce 'quartiles', which are the values exceeded by one quarter and three quarters of the observations, so that one half of a set of observations lies between its quartiles. For the set of observations in *Table 1.1* these would be the 57th and 169th observations; and since there are 167 observations less than 122·5 lb, the first quartile would be taken as $122{\cdot}5 + 1{\cdot}5 \times 5/23 = 122{\cdot}83$ lb while, since there are 38 observations less than 102·5 lb, the second quartile would be taken as $102{\cdot}5 + 18{\cdot}5 \times 5/28 = 105{\cdot}80$ lb. Consequently half the observations in *Table 1.1* lie between 105·80 and 122·83 lb.

If the histogram or frequency diagram of the set of observations is symmetrical, then the quartiles are at the same distance from the median (which is also technically a quartile) and in the above example this is roughly true since $114{\cdot}10 - 105{\cdot}80 = 8{\cdot}30$ lb and $122{\cdot}83 - 114{\cdot}10 = 8{\cdot}73$ lb. These two values provide another indication of the spread, and the average of them, which is $(122{\cdot}83 - 105{\cdot}80)/2 = 8{\cdot}515$, is called the quartile deviation.

* A number of formulae can be used to find the mode more exactly. If the two groups 107·5 – 112·5 and 117·5 – 122·5 are used the mode may be estimated as $115 + (30 - 35)/(2 \times 36 - 30 - 35) \times 2{\cdot}5 = 113{\cdot}21$ lb where 30 and 35 are the frequencies in the two groups and 2·5 is one half of the grouping interval.

The quartile deviation or, as it is frequently called, the probable deviation is roughly equal to two thirds of the standard deviation. The quartiles and quartile deviations are given in *Figures 2* and *3*, and it can be seen that deviations from symmetry are reflected in the unequal distances of the quartiles from the medians.

A similar set of measures is provided by the 'deciles' which are the values exceeded by one tenth, two tenths . . . nine tenths of the observations. Clearly choice of possible measures is large, but we must avoid carrying this idea too far lest the measures cease to be precise and tend to reflect slight sampling irregularities in the observations.

It has been seen that the mean fails as a summary of a set of measurements because it fails to take into account the variability. The standard deviation, which measures this variability, is dependent upon the scale used. Also it is obvious that a standard deviation of, say, 12 lb indicates a relatively greater variability if the mean weight is 40 lb than if it is 120 lb. In consequence a useful measure is provided if the standard deviation is divided by the mean. This gives an index of the relative or proportional variability which is independent of the scale used.

Conventionally this index is multiplied by 100, so that the percentage variability is indicated. This quantity, $100 \times$ (standard devn)/(mean), is called the coefficient of variability or the coefficient of variation. Thus, for the set of measurements in *Table 1.1*, the coefficient of variability is $1{,}238/114{\cdot}58 = 10{\cdot}80$ per cent, while for the two sets of measurements on pp 5, 6, the coefficients of variability are $712/115 = 6{\cdot}19$ per cent and $1{,}231/113 = 10{\cdot}89$ per cent. These values would be unaltered if the measurements had been taken in kg or cwt. However the coefficient of variability is still dependent on the origin, so that different answers would be obtained for the coefficient of variability according to whether, for example, temperature is measured in degrees Centigrade, Fahrenheit or Absolute, although the coefficients would be identical for the Centigrade and Réaumur temperature scales, since the ice point is 0° on both.

1A.13 *Moments and other coefficients*—The variance of a set of observations has been defined on p 6 as the average value of the squared deviations of each observation from the arithmetic mean, or as it is also called, the mean squared deviation. Often as well as the mean squared deviation, the mean cubed deviation, the mean fourth power deviation and so on, are calculated. These latter quantities are known as moments of the third, fourth . . . orders or briefly as the third, fourth, fifth . . . moments. Thus, for example, the set of observations, 104, 107, 111, 112, 114, 115, 117, 119, 124, 127, has an arithmetic mean of 115, and the deviations from this mean are -11, -8, -4, -3, -1, 0, 2, 4, 9, 12, so that the third moment is estimated as

$$[(-11)^3+(-8)^3+(-4)^3+(-3)^3+(-1)^3+0^3+2^3+4^3+9^3+12^3]/10$$
$$=[-1{,}331-512-64-27-1+0+8+64+729+1{,}728]/10=59{\cdot}4$$

and the fourth moment is estimated as

$$[(-11)^4+(-8)^4+(-4)^4+(-3)^4+(-1)^4+0^4+2^4+4^4+9^4+12^4]/10$$
$$=[14{,}641+4{,}096+256+81+1+0+16+256+6{,}561+20{,}736]/10$$
$$=4{,}664{\cdot}4$$

Strictly speaking the appropriate divisor is not 10 since, in the same manner as with the variance, it is necessary to correct for the fact that the true mean is not known exactly if these measures are to be comparable.

Commonly, the third, fourth . . . moments are denoted by μ_3, μ_4 . . . and their estimates by m_3, m_4 . . . , while the variance and its estimate may be denoted alternatively by μ_2 and m_2.

Moments are useful measures since they are independent of the origin chosen. The third moment will be zero if the frequency distribution of the set of observations is symmetrical so that it provides yet another measure of skewness. The greatest use of moments lies however in the ease with which they can be calculated for theoretical frequency distributions, and used as a basis for comparing populations. A fuller discussion of moments must therefore wait until more is known about the types of frequency distributions that are commonly encountered.

1A.14 *Formula for calculation of standard deviation*—In section 1.10 a formula for the calculation of the standard deviation was quoted but not proved. The proof of this formula is not difficult, and since this formula is used in several later chapters it will be proved now.

Suppose $y_1, y_2 \ldots y_n$ are any n quantities with mean $\bar{y}$ then

$$\begin{aligned}\Sigma(y-\bar{y})^2 &= (y_1-\bar{y})^2+(y_2-\bar{y})^2+\ldots+(y_n-\bar{y})^2\\ &= y_1^2-2y_1\bar{y}+\bar{y}^2\\ &\quad +y_2^2-2y_2\bar{y}+\bar{y}^2\\ &\quad +\ldots\ldots\\ &\quad +y_n^2-2y_n\bar{y}+\bar{y}^2\\ &= y_1^2+y_2^2+\ldots+y_n^2-2\bar{y}(y_1+y_2+\ldots+y_n)+n\bar{y}^2\end{aligned}$$

But $\bar{y}=\dfrac{y_1+y_2+\ .\ .\ .\ +y_n}{n}$

$$\therefore\quad \Sigma(y-\bar{y})^2$$

$$=y_1{}^2+y_2{}^2+\ .\ .\ .\ +y_n{}^2-2\frac{(y_1+y_2+\ .\ .\ .\ +y_n)}{n}(y_1+y_2+\ .\ .\ .\ +y_n)$$

$$+n\frac{(y_1+y_2+\ .\ .\ .\ +y_n)^2}{n^2}$$

$$=y_1{}^2+y_2{}^2+\ .\ .\ .\ +y_n{}^2-2\frac{(y_1+y_2+\ .\ .\ .\ +y_n)^2}{n}+\frac{(y_1+y_2+\ .\ .\ .\ +y_n)^2}{n}$$

$$=y_1{}^2+y_1{}^2+\ .\ .\ .\ +y_n{}^2-\frac{(y_1+y_2+\ .\ .\ .\ +y_n)^2}{n}$$

$$=\Sigma y^2-\frac{(\Sigma y)^2}{n}$$

If we now put $y_1=x_1-a,\ \ y_2=x_2-a\ .\ .\ .\ y_n=x_n-a,$ then $\bar{y}=\bar{x}-a$ and $y_1-\bar{y}=x_1-\bar{x},\ \ y_2-\bar{y}=x_2-\bar{x}\ .\ .\ .\ y_n-\bar{y}=x_n-\bar{x}.$

Thus

$$\Sigma(x-\bar{x})^2=(x_1-\bar{x})^2+(x_2-\bar{x})^2+\ .\ .\ .\ +(x_n-\bar{x})^2$$

$$=(y_1-\bar{y})^2+(y_2-\bar{y})^2+\ .\ .\ .\ +(y_n-\bar{y})^2$$

$$=y_1{}^2+y_2{}^2+\ .\ .\ .\ +y_n{}^2-\frac{(y_1+y_2+\ .\ .\ .\ +y_n)^2}{n}\text{ by the above formula}$$

$$=(x_1-a)^2+(x_2-a)^2+\ .\ .\ .\ +(x_n-a)^2$$

$$-\frac{[x_1-a+x_2-a+\ .\ .\ .\ +x_n-a]^2}{n}$$

$$=\Sigma(x-a)^2-\frac{[\Sigma(x-a)]^2}{n}\text{ as required.}$$

SUMMARY OF PP 13 TO 18

When a set of measurements is not symmetrically distributed about the mean, various other measures of central tendency may be used. The median and mode are two possible measures, and the differences (mean – median)/(standard devn) and (mean – mode)/(standard devn) indicate the extent of the deviation from symmetry. The first of these two measures of skewness is approximately one third of the second. Further useful measures are provided by the quartiles, which are the values exceeded by exactly one quarter and three quarters of the observations. The difference between these indicates the spread of the frequency distribution, while the skewness is further indicated by the comparative distances of the quartiles from the median.

The coefficient of variability or standard deviation divided by the mean may be used to compare the relative variation of different samples. This index is independent of the scale used, but is dependent on the origin of the scale.

EXAMPLES

6 The frequency distribution of weights of 7,749 adult males as reported by the British Association Anthropometric Committee is:

Weight lb	90-	100-	110-	120-	130-	140-	150-	160-	170-	180-	
Frequency	2	34	152	390	867	1,623	1,559	1,326	787	476	
Weight lb	190-	200-	210-	220-	230-	240-	250-	260-	270-	280-	
Frequency	263	107	85	41	16	11	8	1	0	1	*Total*: 7,749

Construct a histogram to demonstrate the distribution and show that the mean, median and quartiles are 157·2, 155·2, 143·0, and 168·9 lb. Using a value of 20·4 lb for the standard deviation, show that the skewness of the distribution is comparable in magnitude to the skewness in *Figure 3* but in the opposite direction.

7 The cephalic indices (100 times the ratio of head length to head breadth) of 756 Aberdeen children less than five years old were:

Mean index	70	72	74	76	78	80	82	84	86	88	90	
Number of children	1	2	34	107	183	205	140	50	21	11	2	*Total*: 756

Construct a histogram and frequency diagram to demonstrate the distribution and show that the mean, median and standard deviation are 79·59, 79·50 and 3·00. Hence show that the skewness of the distribution is negligible and verify this using the quartiles.

8 In a test of ability in English the scores of 2,772 pupils were:

Score	50-	60-	70-	80-	90-	100-	110-	120-	130-	140-	
Frequency	11	48	196	403	636	693	472	212	95	6	*Total*: 2,772

Show that the mean score is 101·1 and that the standard deviation is 15·8. Use the median to show that the scores are symmetrically distributed about their mean.

9 The age distributions of Bristol mothers at *primiparae* in 1932 and 1937 were:

Age	14-	16-	18-	20-	22-	24-	26-	28-	
Frequency 1932	2	22	113	235	302	378	362	235	
Frequency 1937	1	30	122	312	412	409	392	309	
Age	30-	32-	34-	36-	38-	40-	42-	44-	
Frequency 1932	170	103	61	30	28	6	3	1	*Total*: 2,051
Frequency 1937	213	141	87	54	18	9	4	4	*Total*: 2,517

Draw frequency diagrams to demonstrate the distribution and show that the mean, median and standard deviation are 26·28, 25·86 and 4·63 in 1932, and 26·37, 25·87 and 4·75 in 1937.

The difference between the distributions is not very large, since the means and standard deviations differ by about a month, although when two thousand cases are considered this amounts to a total difference of about 200 years. What has to be decided from these figures is whether this difference is a purely chance difference arising from the natural variability in the ages at *primiparae,* or whether sufficient cases have been considered to rule out the possibility that this is a chance difference and to conclude that a change of social significance has occurred during the five year period. This question must remain unanswered until the next chapter when the accuracy of the mean of a set of observations will be considered.

10 The estimated distribution of annual income among the unmarried women of a certain city was:

Income	£0-	£50-	£100-	£150-	£200-	£250-	£300-	
Frequency	13	189	671	584	313	180	67	
Income	£350-	£400-	£450-	£500-	£550-	£600-	£650-	
Frequency	31	23	18	13	9	5	6	*Total*: 2,122

Construct a histogram and frequency diagram to illustrate the distribution, and calculate the mean, median, quartiles and standard deviation of incomes.

2

NORMAL DISTRIBUTION

2.1 *Frequency distributions*—At the beginning of the previous chapter it was explained that it is convenient to regard a set of measurements as a sample from a population of possible measurements, although the population may not exist. For example, today's weather may be regarded as a sample of the population of possibilities, among which would be included days with, say, mean temperatures 57° F, 1 in rainfall and 5 hr of sunshine, although this particular combination may not have been experienced. It is likewise convenient to consider the frequency distribution of a set of observations as arising from a frequency distribution in the population giving the true proportions in each grouping interval. Thus, the frequency distribution of children's weights would give the proportion of children with weights between, say, 112·5 and 117·5 lb in the population, and the proportion in this group in *Table 1.1*, *i.e.* 36/225 = 0·16, is regarded as an estimate of the proportion in the whole population. Since a set of measurements is normally taken in order to determine the character of the population, it is the frequency distribution of the population which we attempt to estimate in general.

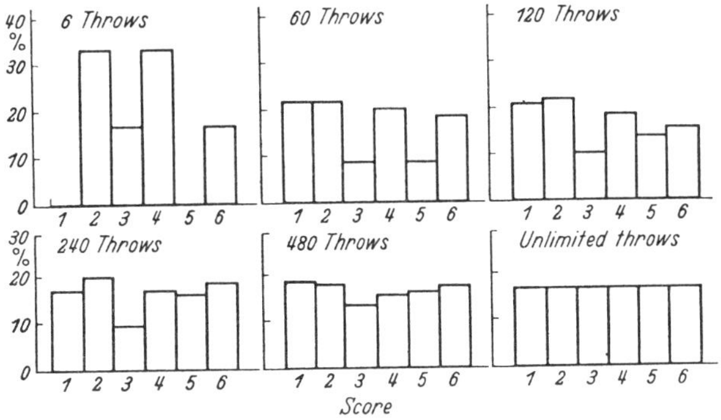

Figure 4. Histograms of the scores observed in throwing a die

To consider a simple example, suppose a die is thrown a number of times and the numbers of ones, twos, threes . . . cast are recorded, then the relative frequencies of each can be shown in a histogram as in *Figure 4*. As the number of throws is increased so is the regularity of the histogram, and the frequency distribution of the sample is said to tend to the frequency distribution of the population. Here the frequency distribution of the population is 16·7 per cent ones, 16·7 per cent twos, 16·7 per cent threes . . . a fact which could be predicted from the six sided form of the die. It should be noted that an *a priori* knowledge of the distribution enables us to predict with reasonable accuracy the behaviour of a sample. This fact is important since it will be seen later that it is sometimes possible to find the distribution of the population from a small sample, and to use this knowledge to predict the behaviour of a large sample. Thus, if the frequency distribution of children's weights can be found from, say, twenty observations, this can be used to predict the proportion of children with weights between any two values.

To demonstrate further the frequency distribution of the population, the total score when six dice were thrown simultaneously was recorded. This score cannot be less than six nor more than thirty six, and the frequency distribution of the population can again be determined using the laws of probability. For example, the proportion of thirty sixes *i.e.* six sixes will be

$$\frac{1}{6} \times \frac{1}{6} \times \frac{1}{6} \times \frac{1}{6} \times \frac{1}{6} \times \frac{1}{6} = \left(\frac{1}{6}\right)^6 = \frac{1}{46,656}$$

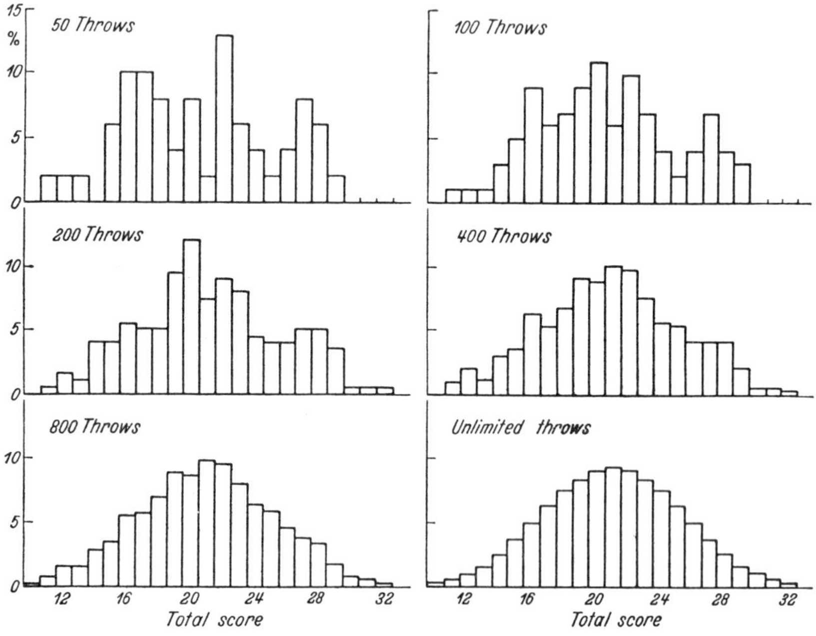

Figure 5. *Histograms of the total scores observed in throwing six dice*

Figure 5 shows how as the number of throws of the six dice is increased the histograms become more regular and tend to the frequency distribution of the population. It should be remembered that the proportion of occasions on which the score is between any two values is represented by the area between those values. Here, in the vast majority of throws (99·1 per cent to be exact), the score will be between eleven and thirty one.

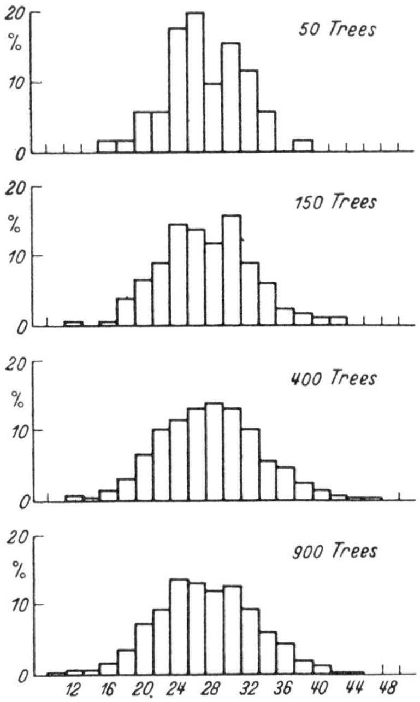

Figure 6. Histograms of tree girths

A more practical example is shown in *Figure* 6 which gives sets of tree girths measured by forestry students. As the size of sample is increased the irregularities in the histogram disappear and the form of frequency distribution of the population becomes more apparent. Here the distribution is not determined *a priori*, as with the distribution of scores of dice. It is however noticeable that the distributions in *Figures* 5 and 6 bear marked similarities. It will be seen later that this similarity is not purely coincidental but is the consequence of a law of variability which determines the shape of many frequency distributions experienced in practice.

2.2 *Frequency curves*—In *Figures* 4 and 5, examples are considered where the final measurements must be whole numbers, since fractions are not indicated on a die. Such cases do occur in practice when counts of say, insects, plants or animals are being made, but it is more usual for measurements to be able to take all values between two limits and not only whole numbers. For example, all weights between 85 and 145 lb are possible since children's weights do not increase by jumps of one pound. Measurements such as these are said to be continuous.

Grouping was introduced in Chapter 1 as a method of comparing the relative frequencies of different measurements. In order to make the comparison accurate it is necessary to choose the grouping interval sufficiently large to include several measurements. If, however, a large number of measurements is made the grouping interval may be fairly small, and when the number of observations is increased indefinitely, so that the frequency distribution of the population is obtained, the grouping interval can be made as small as desired. *Figure* 7 shows how it is possible to reduce the grouping interval of the histogram of a frequency distribution so that ultimately a curve is obtained instead of a set of blocks. This process of reducing the grouping interval indefinitely gives a frequency curve.

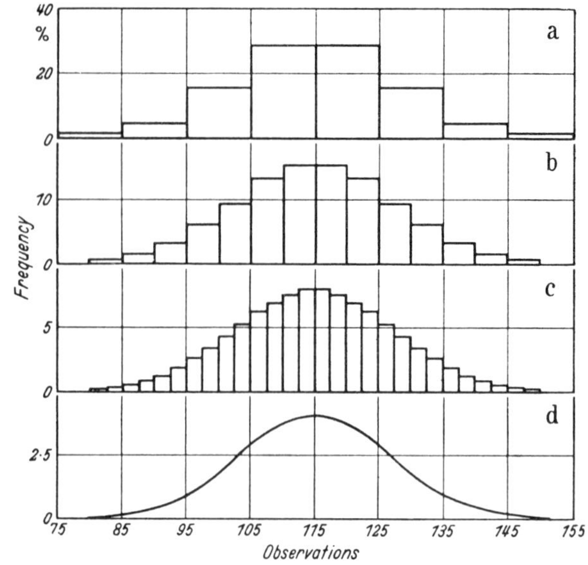

Figure 7. Frequency curve obtained from histogram

Frequency curves retain the property of histograms that the area between any two values gives the proportion of observations between the two values, while the relative frequencies of observations are represented by their distances from the axis. The most frequent observation (or mode) thus corresponds to the highest point or maximum of the frequency curve. For example, in *Figure 7*, the weight of 100 lb is observed roughly three times as often as the weight of 90 lb, while approximately one out of every eight observations is less than 100 lb.

2.3 *Normal distribution*—One of the requirements of a summary form of a set of observations is that it must be possible to find the proportion of observations between any two values. The forester will require from a summary the proportion of trees that can be used for a certain purpose *i.e.* with girths between any two particular values; the psychologist wants the percentage of the population exceeding a certain score; and generally the summary will be expected to convey information on the distribution of observations. If it does not, it fails in an important aspect. Fortunately, specification of the distribution is often not very difficult.

It can be shown mathematically that whenever a measurement is the sum total of a large number of small independent effects, no one of which predominates, the distribution of observations will take the same form. This form is called the normal distribution and a knowledge of the origin and scale determines it completely *i.e.* once the mean and spread are known all else is determined. This is not generally true of distributions, since the mean and spread convey no information, for example, on the asymmetry of the distribution, but the normal distribution is symmetrical and all of its properties can be found from its mean and spread. *Figure 7* gives, in fact, the frequency curve of the normal distribution, and the same curve is shown in *Figure 8* with different spreads. That this curve is not a mere mathematical abstraction has been demonstrated by numerous practical applications, some of which are given in examples *8* and *9* and *Figures 5, 6, 9, 10* and *11*.

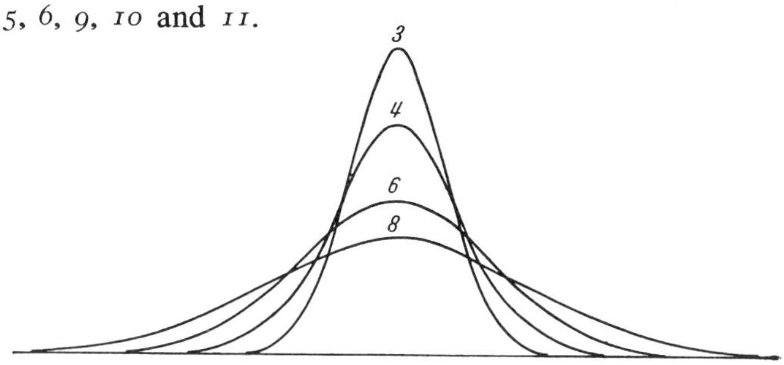

Figure 8. Normal curves with relative spreads of 3:4:6:8

It is interesting to postulate some of the many causes which have contributed jointly to the measurements in each of these applications. For instance, the yields of wheat and the girths of trees will have resulted partly from variations in soil fertility, humidity, temperature, activity of pests, competition, and other environmental factors and partly from a multitude of genetical factors. Likewise, most of the other measurements result from a complex of environmental and genetical factors. The exceptions, here, are the distribution of observations of the right ascension of the Pole Star which is caused by a series of small errors in measurement, and the distribution of the total score on six dice in which the individual scores on the dice act as six independent effects contributing to the final measurement.

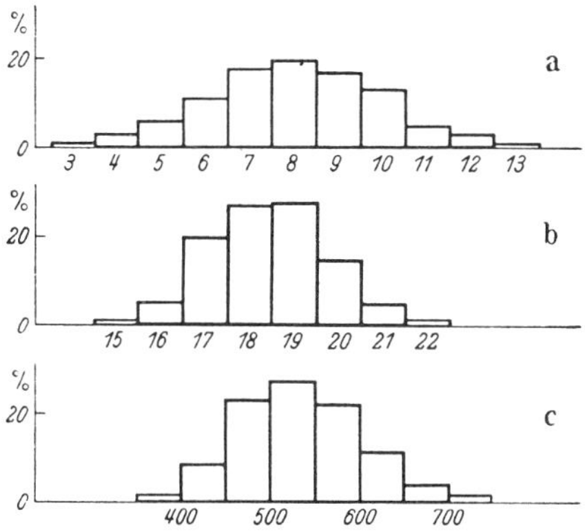

Figure 9. a *Frequency distribution of the sizes of 2,020 litters from Duroc-Jersey pigs [Data of* PARKES, A. S. *Biometrika* 15 (1923) 373]; b *Frequency distribution of heights of 917 infants at birth (loc. cit.);* c *Frequency distribution of yields from 700 wheat plots*

The normal distribution cannot be applied to every measurement and the reasons why this is so are worth noting. First, there may not be a large number of effects, and one particular effect may predominate, as with throws of a single die shown in *Figure 4*. Secondly, the effects may not be independent, as with rainfall and temperature, but in general this will not affect the distribution provided a large number of effects operate on the measurements. Thirdly, the effects may not sum, which is a more likely cause of a deviation from the normal form.

Thus a 10° rise in temperature may cause a proportional rather than an absolute change in a measurement. It is also true that weight measurements tend to be affected in a proportional rather than absolute fashion, so the measurement instead of resulting from the sum of a large number of effects results from the product of these effects. This difficulty can often be overcome by using the logarithm of the measurement since if $x = a \times b \times c \times \ldots$ then $\log x = \log a + \log b + \log c + \ldots$ and is determined by the sum of the effects.

It should be remembered that a great many observations have to be taken before any deviation from normality can be detected, since natural variability tends to hide the shape of the distribution. This is demonstrated by the initial irregularities in *Figures 4* and *5*, and the reader should not be misled into believing that the regularity of *Figures 9, 10* and *11* will be achieved with a small number of observations.

It is often possible to assume that a set of observations is normally distributed as a result of previous experience with similar measurements. If it is known that similar factors will affect both sets of measurements, although possibly to different degrees, the multitude of factors causing 'normality' for one set of observations will usually cause it for the other. The normal distribution is thus commonly experienced in practice and during the next few chapters the applications of this distribution will be considered. This does not rule out the possibility of encountering non-normal distributions. However, the methods which will be

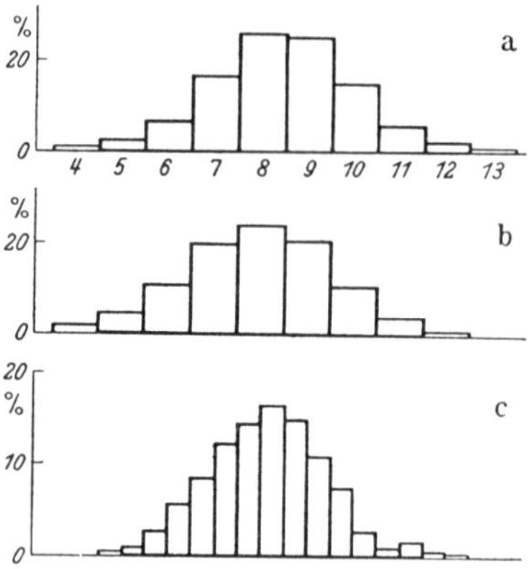

Figure 10. a *Frequency distribution of the weight of seed from 1,668 garden beans* [*Data of* HARRISS, J. A. *Biometrika* 9 (1913) 436]; b *Frequency distribution of size of class in 34,005 primary and secondary schools;* c *Frequency distribution of the ratio of breadth to length of corolla in 1,000 foxgloves* [*Data of* WARREN, E. *Biometrika* 11 (1916) 303]

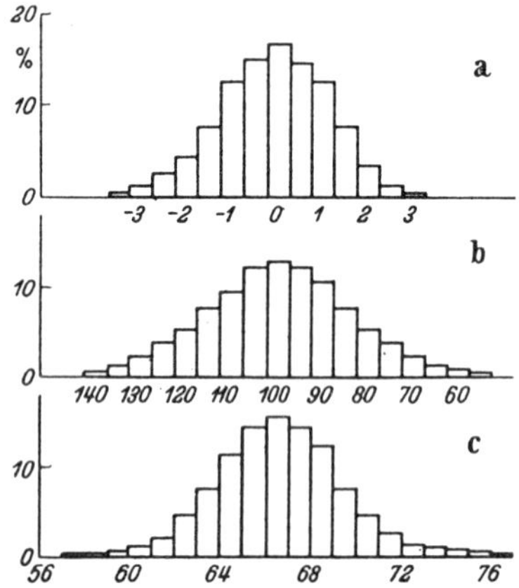

Figure 11. a *Frequency distribution of 1,000 observations of the right ascension of the Pole Star* [*Data of* WHITTAKER, Sir E. *and* ROBINSON, G. *Calculus of Observations* fourth edition, p 174, London, 1944]; b *Frequency distribution of the intelligence quotients of 5,562 11-year old children;* c *Frequency distribution of height of 8,585 adult males* [*Final Report of Anthropometric Committee to British Association* (1883) 256]

given are insensitive to deviations from normality and may be applied to distributions which are not markedly non-normal. Further it will be seen later that the methods derived for the normal distribution can be applied to sets of observations which are not normally distributed by transforming the scale of measurement.

2.4 *Applications of normal distribution*—As has been stated in the last section, the normal distribution is determined by its mean and spread, from which it is possible to derive any other information that is required. For example, the mean deviation is 0·7979 times the standard deviation for the normal distribution. If we choose the origin at the mean of the distribution and the scale of the measurements is made equal to the standard deviation, all normal distributions are equivalent. This is demonstrated graphically in *Figure 12* which shows the equivalence of three distributions with different standard deviations. This same statement may be expressed mathematically as follows: if μ is the mean and σ is the standard deviation, then the proportion of measurements exceeding $\mu+\sigma d$ depends only upon d *i.e.* the deviation from the mean expressed in terms of the standard deviation. Suppose, for example, the mean and standard deviation of the girths of a group of trees are 45 and 10 in respectively, then the proportion of trees with girth exceeding, say, 57 in depends only upon $(57-45)/10=1{\cdot}2$, which is the deviation from the mean expressed in terms of the standard deviation. This quantity is called a normal deviate. The proportion of observations exceeding any normal deviate can be found by referring to *Table I* at the back of the book*. The

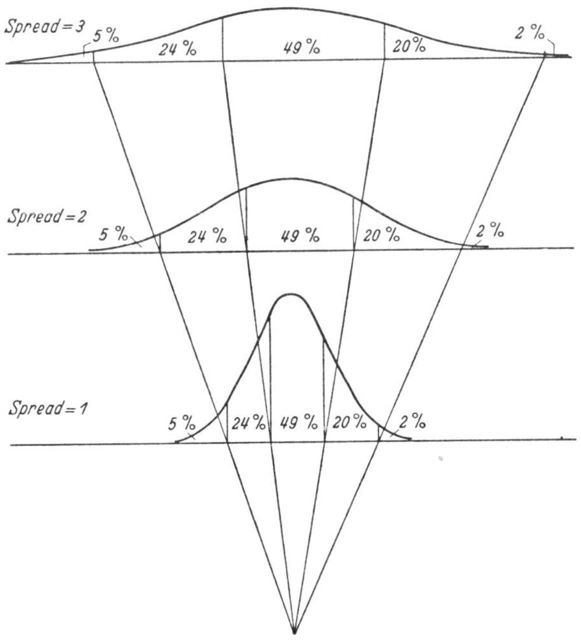

Figure 12. Normal distributions with different spreads

* This table is derived by calculating the area under the normal curve lying beyond d.

use of this table gives the required percentage, so that 11·51 per cent (corresponding to a normal deviate of 1·2) of the girths would exceed 57 in. This percentage might be compared with the value obtained directly from example *1*, in which the number of trees with girths exceeding 57 in may be taken as $\frac{1}{2}\times 52+28+6+2=62$, which is 13·48 per cent of the total number.

It is worth while demonstrating more extensively the uses of this table by further examples.

a The exact values of the mean and standard deviation from example *1* might be used to calculate the proportion of girths exceeding 57 in. The difference between the mean and 57 in is 12·49, and the normal deviate is 12·49/10·78 = 1·16. Using *Table I*, the required percentage is 12·75, which agrees quite well with the observed value of 13·48 per cent. It should be noticed that the difference of 0·73 per cent in these two values represents a difference of only 3 trees in a sample of 460, and that as the size of the sample is increased the former value is likely to prove more accurate.

b The mean and standard deviation in a psychological test of ability in English were 101·1 and 15·8. Suppose it is required to calculate the percentage with scores between 80 and 120.

The differences between these scores and the mean are −21·1 and 18·9, which when expressed in terms of the standard deviation are −21·1/15·8 = −1·34 and 18·9/15·8 = 1·20. It is then required to find the percentage of observations between the normal deviates −1·34 and 1·20. Now, from *Table I*, −1·34 is exceeded by 90·96 per cent and 1·20 is exceeded by 11·51 per cent so that 90·96 − 11·51 = 79·45 per cent of observations must lie between these values. Thus, in a sample of 2,772 pupils, 2,772 × 0·7945 = 2,202 would be expected to have a score between 80 and 120. This value might be compared with the value of 2,204 actually observed in example *8*. In view of the variability in the scores this agreement between the two values is remarkable, but it serves to emphasize the fact that the mean and standard deviation conveniently and concisely summarize a set of observations.

c The red blood cell counts on twenty 2-week old piglets in millions per cm^3 were 4,300, 5,200, 5,300, 5,800, 6,000, 6,300, 6,400, 6,600, 6,700, 6,900, 7,100, 7,200, 7,400, 7,500, 7,800, 8,000, 8,400, 8,700, 9,100, 9,300, from which it is required to find the percentage of piglets with counts over 9,000. Of the 20 piglets in the sample, 2, or 10 per cent, have counts exceeding 9,000, but this percentage will not be accurate since it is based upon a very small sample. The counts are however fairly symmetrically placed about the mean of 7,000, and since previous observations have shown that blood counts are usually normally distributed it is possible to make use of the normal distribution to obtain a better estimate of the correct percentage.

The sum of squares of deviations from the mean is $2{,}700^2+1{,}800^2+1{,}700^2+\ldots+2{,}300^2=33{,}620{,}000$. The estimated variance is 33,620,000/19 = 1,769,474 and the standard deviation is 1,330. The normal deviate is thus calculated as (9,000 − 7,000)/1,330 = 1·50, which is exceeded by 6·68 per cent of the observations. This percentage is still subject to error since neither mean nor standard deviation is exactly determined, but it is likely to be more accurate than the crude value of 10 per cent.

d Suppose it is known that 7,000 and 1,200 are accurate estimates of the mean and standard deviation of the blood counts of 2-week old piglets, and a piglet fed on a special diet has a blood count of 10,600. Then this high value indicates the probable effect of the diet, but the possibility of this figure occurring by chance cannot be ruled out. To test this we might find the percentage of observations that would normally exceed 10,600 by using a normal deviate (10,600 − 7,000)/1,200 = 3·0 in *Table I*. This indicates that as high a value would normally occur in only 0·135 per cent of observations, which means that, if the diet has had no effect, the value observed would occur less than one in seven hundred times by pure chance. We thus have to conclude either that the diet has had an effect or that a very unlikely observation has been taken. In general, we would conclude that the diet had affected the blood count, but the possibility of this being a chance effect is not completely ruled out.

This latter example is rather trivial since several observations are usually required to detect any differences and often experimental differences are relatively small. However, it serves to indicate the line of reasoning which will be followed in testing treatment effects.

On the assumption that there are no treatment effects, we find the percentage of occasions on which as extreme a set of values as the observations would be obtained. If this percentage is small it indicates a set of observations would not usually occur by pure chance. We then conclude that the treatments have been effective. Nevertheless there is always a possibility, however small, that the effects are chance effects.

The assumption that the treatments have had no effect is called the null hypothesis. If the percentage worked out on the null hypothesis is small, the hypothesis is rejected and the effectiveness of the treatments is concluded. The process here might be compared with the common method of employing a hypothesis or theory until observations are taken which are so unlikely under the hypothesis that it has to be rejected. Commonly, the term statistical induction is applied to this process. In the following chapters many examples will be given of its application.

2.5 *Alternative form of normal-deviate table*—When the mean and standard deviation of a set of observations are known it is often useful to know the region within which, say, nine tenths of the observations lie. This can be found from *Table I.* Since 95 per cent of observations have a deviate exceeding $-1{\cdot}64$ and 5 per cent have a deviate exceeding $1{\cdot}64$, 90 per cent or nine tenths of the observations have deviates between $-1{\cdot}64$ and $1{\cdot}64$. Similarly, the limits within which any percentage of the observations lies can be found but, since these limits have to be calculated from *Table I,* it is convenient to use an alternative form of this table. Such a form is shown in *Table II* of the Appendix, which gives the limits corresponding to various percentages. Thus, if μ is the mean and σ the standard deviation, 50 per cent of the observations lie between $\mu - 0{\cdot}66\,\sigma$ and $\mu + 0{\cdot}66\,\sigma$ and 99 per cent lie between $\mu - 2{\cdot}58\,\sigma$ and $\mu + 2{\cdot}58\,\sigma$. It will be seen later that this is a convenient form of tabulation for many purposes.

SUMMARY OF PP 21 TO 29

It has been shown that one particular distribution—the normal distribution—occurs very frequently in practice. This distribution arises when the measurement is determined by the sum of a large number of small effects and it depends only upon its mean and standard deviation which can be used to find any other characteristic of the distribution.

The percentage of observations exceeding any value is determined by the normal deviate, which is the value reduced by the mean and divided by the standard deviation *i.e.* $(x-\mu)/\sigma$. Corresponding values of normal deviates and percentages have been tabulated so that one can be found from the other.

EXAMPLES

11 Using the following table of ordinates, plot the normal curve with zero mean and unit standard deviation. Use this curve to show that roughly one sixth of the observations have a normal deviate exceeding one.

Normal deviate	0	±0·2	±0·4	±0·6	±0·8	±1·0	±1·2
Ordinate of normal curve	0·40	0·39	0·37	0·33	0·29	0·24	0·19
Normal deviate	±1·4	±1·6	±1·8	±2·0	±2·2	±2·4	±2·6
Ordinate of normal curve	0·15	0·11	0·08	0·05	0·035	0·02	0·01

12 The mean and standard deviation of tree girths shown in *Figure 6* are 27·13 and 5·60, show that the normal deviate corresponding to a girth of 36 in is 1·58, and hence conclude that 5·72 per cent of trees have a girth exceeding 36 in. This value should be compared with the actually observed percentages in groups of 50, 150, 400 and 900 trees, which were 2·00, 6·67, 7·38, and 6·29 per cent.

13 The mean death rate in 171 Scottish burghs in 1937 was 11·91 per thousand, and the standard deviation was 2·52. Verify that 22·5 per cent of burghs would be expected to have a death rate of 10 per thousand or less, and 11·0 per cent would be expected to have a death rate exceeding 15 per thousand.

14 The level of resistance of animals to disease is measured by their antibody level. Factors affecting this level tend to cause proportional rather than absolute changes, so that it is the logarithm of the antibody level which is normally distributed and which is used in the analysis of experiments. The mean and standard deviation of this measure for a flock of sheep were 0·78 and 0·45. If 1·50 is regarded as a satisfactory level of immunity, show that about 5 per cent of the sheep have reached this level.

15 The frequency distribution of the weights given in example *6* is not symmetrical and obviously not normal. This is due to weight changes being proportional rather than absolute and so the logarithm of the weights should be used.

Construct a frequency table using the logarithm of the weight in the form

Log weight	1·954-	2·000-	2·041- . . .
Frequency	2	34	152 . . .

and demonstrate it by a histogram as shown in *Figure 13*.

When the logarithm of the weight is used, the grouping interval varies and the area of each rectangle should be chosen proportional to the frequency. The symmetry of the histogram is rather difficult to assess as a consequence of the irregular grouping interval. Thus if the logarithm of an observation is to be used it is preferable to choose the grouping so that the logarithms are equally spaced. For example, if the grouping for the logarithms is chosen as 1·96-2·00-2·04-2·08 . . . the grouping for the original weights should be taken as 91·2-100·0-109·6-120·2- . . . .

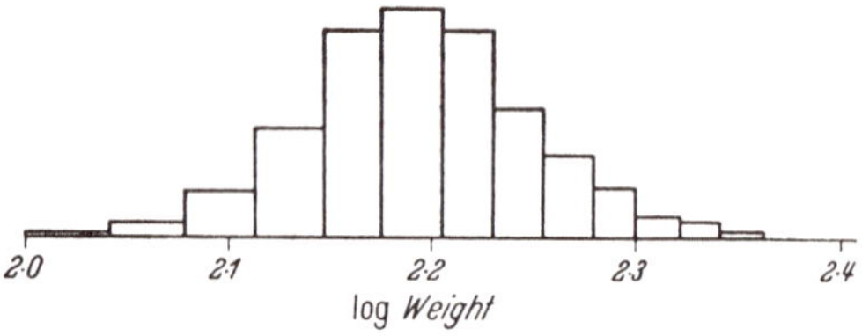

Figure 13. Frequency distribution of the logarithms of the weights of 7,749 adult males

EXTENDED DEVELOPMENT

2A.6 *Theoretical distributions: the binomial*—If a coin is tossed it is equally likely to come down heads or tails. This does not mean that if the coin is tossed twice it will necessarily come down heads once and tails once, but only that about one half of a series of tosses will be heads. An alternative statement of the same fact is that the probability of a head being tossed is one half. This distribution, which might be compared with the distribution in *Figure 4*, can be deduced from the form of the coin and is the simplest type of theoretical frequency distribution.

Suppose two coins are tossed instead of one, then there are four equally likely possibilities; heads may be obtained on both coins; a head may turn up on the first coin and a tail on the second; a tail may turn up on the first coin and a head on the second; or two tails may turn up.

Thus of the four possibilities, two give one tail and one head, corresponding to the two ways of obtaining one head and one tail on two coins. As a result, if two coins are tossed the probability of two heads turning up is one quarter, of one head and one tail is one half, and of two tails is one quarter *i.e.* if the two coins are tossed a large number of times, two tails will turn up in roughly one quarter of the trials.

The same argument might be used to find the theoretical distribution for the tosses of three coins. There are eight equally likely possibilities given in *Table 2.1*:

	First coin	*Second coin*	*Third coin*	*No. of heads*
1	Head	Head	Head	3
2	Head	Head	Tail	2
3	Head	Tail	Head	2
4	Tail	Head	Head	2
5	Tail	Tail	Head	1
6	Tail	Head	Tail	1
7	Head	Tail	Tail	1
8	Tail	Tail	Tail	0

Hence the probabilities of getting three, two, one or no heads are one eighth, three eighths, three eighths or one eighth respectively, and if three coins are tossed a large number of times three tails will turn up in roughly one eighth of the trials.

More generally, if n coins are tossed, there are $2 \times 2 \times \ldots \times 2 = 2^n$ equally likely possibilities, and exactly r heads occur in $\dfrac{n(n-1)\ldots(n-r+1)}{1 \times 2 \ldots r}$ of these. The probability of r heads being tossed is thus

$$\frac{n(n-1)\ldots(n-r+1)}{1 \times 2 \ldots r} \quad \frac{1}{2^n}$$

Using this value it is possible to build up theoretical distributions. For example if 10 coins are tossed the probabilities of getting 0, 1, 2, ... heads are:

$$\frac{1}{2^{10}}, \quad \frac{10}{1} \times \frac{1}{2^{10}}, \quad \frac{10 \times 9}{1 \times 2} \times \frac{1}{2^{10}}, \ldots\ldots\ldots\ldots$$

i.e.

$$\frac{1}{2^{10}}, \frac{10}{2^{10}}, \frac{45}{2^{10}}, \frac{120}{2^{10}}, \frac{210}{2^{10}}, \frac{252}{2^{10}}, \frac{210}{2^{10}}, \frac{120}{2^{10}}, \frac{45}{2^{10}}, \frac{10}{2^{10}}, \frac{1}{2^{10}}$$

This theoretical distribution is shown with several others in *Figure 14*. It is immediately noticeable that as the number of coins is increased the distribution tends to the normal form, but this is not really surprising since the number of heads is determined by the sum of the results from each coin. Thus, if a coin is tossed ten times, say, the number of heads turning up will tend to be normally distributed about the mean number of heads

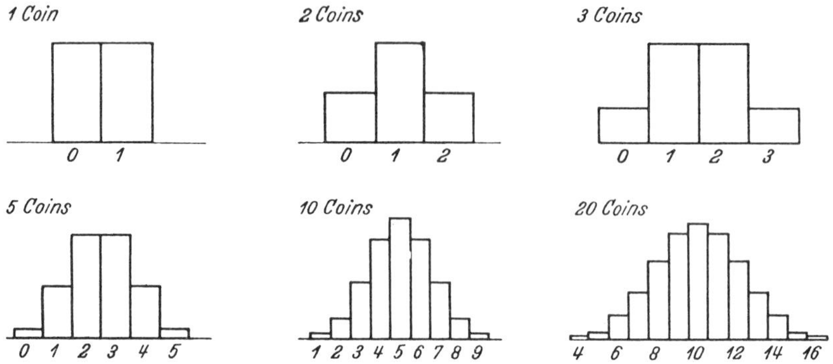

Figure 14. Frequency distribution of the numbers of heads turning up in coin tossing

i.e. five. This, in itself, is of no practical importance, but it is possible to make use of a generalization of this.

If the probability of a head turning up is $\frac{1}{3}$, or $\frac{1}{4}$, or generally p, then we may find the probability of getting r heads as a result of n tosses. This probability is given by the general binomial distribution, which is expressed in the following statement. If p is the probability of success in a single trial then the probability of r successes in n trials is

$$\frac{n(n-1)\dots(n-r+1)}{1 \times 2 \dots r} p^r (1-p)^{n-r}$$

We may then derive frequency distributions of the numbers of heads such as are shown in *Figure 15*. These distributions all tend to normality, and it is possible to predict the means and standard deviations of these limiting normal distributions. This gives the following theorem: If p is the probability of success in a single trial the number of successes in n trials tends to be normally distributed with mean np and standard deviation $\sqrt{[np(1-p)]}$ *i.e.* the variance equals $np(1-p)$. The consequences and applications of this theorem will be considered in the next section.

2A.7 *Application of normal approximation to binomial distribution*—In tossing a coin we know that although the probability of getting a head is one half, we shall not in general get exactly one half of the tosses as

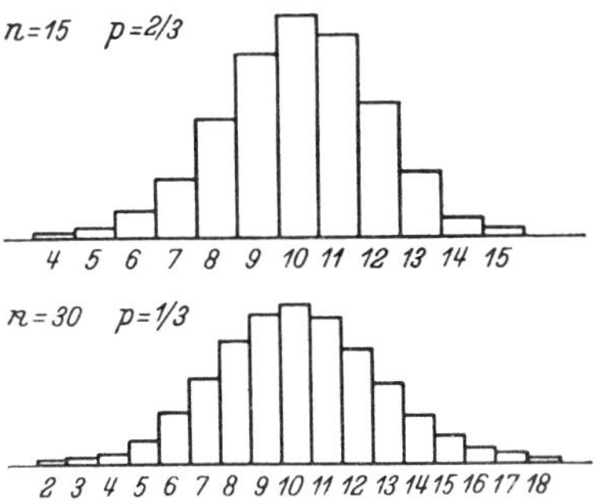

Figure 15. Frequency distribution of the numbers of successes in n trials, where the probability of success in a single trial is p

heads. The theorem in the last section indicates the extent to which the observed proportion is likely to deviate from the true proportion, and this will prove useful both in finding limits for the observed proportion when the true proportion is known and *vice versa.*

Three examples will demonstrate the applications of this theorem:

a If at a given stage of the breeding season one half of chicks hatched for a certain breed are cocks, how many eggs must be hatched to ensure that at least 20 pullets are obtained?

It is obvious that we can never be absolutely certain of getting 20 pullets, but a high degree of certainty can be obtained by increasing the numbers sufficiently. Thus if 40 eggs are hatched, we are as likely to get less than 20 pullets as we are to get more than 20, so that more than 40 eggs would be required for any degree of certainty. If 50 eggs are used then the mean and standard deviation of the number of pullets hatched are 25 ($=50\times\frac{1}{2}$) and 3·54 [$=\sqrt{(50\times\frac{1}{2}\times\frac{1}{2})}$] so that the normal deviate corresponding to 20 pullets is $(20-25)/3{\cdot}54=-1{\cdot}41$ which, from *Table I,* is exceeded in 92 per cent of trials. This shows that the use of 50 eggs makes it likely that more than 20 pullets will be hatched, but in 8 per cent of trials the required number will not be achieved. If the number of eggs is increased to 60, then the mean and standard deviation become 30 ($=60\times\frac{1}{2}$) and 3·87 [$=\sqrt{(60\times\frac{1}{2}\times\frac{1}{2})}$], and the normal deviate is now $(20-30)/3{\cdot}87=-2{\cdot}58$. In this instance, more than 20 pullets will be obtained in 99·5 per cent of trials, and less than 20 pullets will hatch in 0·5 per cent of trials. This degree of certainty would usually be sufficient since only once in two hundred times would the number fall short of requirements. If a higher degree of certainty were needed a correspondingly greater number of eggs would be required.

The number of eggs required for any degree of certainty can be obtained by a trial and error process, or by a reversal of the above process. For example, if we want to ensure that more than 20 pullets will hatch in 99·9 per cent of trials, the normal deviate is $-3{\cdot}1$ and this is equal to $(20-\frac{1}{2}\times n)/\sqrt{(n\times\frac{1}{2}\times\frac{1}{2})}$. This gives rise to the equation

$$(20-\tfrac{1}{2}n)/\sqrt{(\tfrac{1}{4}n)}=-3{\cdot}1$$

$$(20-\tfrac{1}{2}n)^2=(3{\cdot}1)^2\tfrac{1}{4}n$$

$$n^2-89{\cdot}61n+1{,}600=0$$

$$n=65{\cdot}0 \text{ or } 24{\cdot}6$$

so that 65 eggs would be required. If only 25 eggs are used less than 20 pullets will be hatched in 99·9 per cent of trials, and this corresponds to the other solution of the given equation.

b In genetical work it is often possible to predict the proportions of animals or plants with certain characteristics arising from particular conditions. For example, under the Mendelian hypothesis the crossing of a tall race of peas with a dwarf race should yield in the second generation tall and dwarf peas in the ratio 3:1. In testing this, it is necessary to ensure that any deviations from the exact 3:1 ratio are chance deviations and are not due to the influence of another genetical factor (as occurs in linkage). For example, J. G. MENDEL in his original experiments got 787 tall and 277 dwarf plants from 1,064 crosses, which is 74 per cent tall and 26 per cent dwarf, from which it was necessary to decide whether the deviations were due to chance or not. The expected number of dwarf plants is $\frac{1}{4}\times 1{,}064=266$, and this has a standard deviation of $\sqrt{(1{,}064\times\frac{1}{4}\times\frac{3}{4})}=14{\cdot}1$. Thus the normal deviate corresponding to 277 is $(277-266)/14{\cdot}1=0{\cdot}78$, and this is exceeded by pure chance in 21·8 per cent of samples, so that the deviations from the theoretical 3 · 1 ratio can be attributed to chance.

c Suppose 60 per cent of the population of a town approves a certain motion. If an investigator interviews 600 people chosen at random, then the number of people expected to approve the motion is 360 and this has a standard deviation of $\sqrt{(600 \times 0{\cdot}60 \times 0{\cdot}40)} = 12$. Thus, from *Table II*, in 50 per cent of samples the number approving will lie between $360 - 0{\cdot}67 \times 12$ and $360 + 0{\cdot}67 \times 12$ *i.e.* between 352 and 368, and in 95 per cent of samples between $360 - 1{\cdot}96 \times 12$ and $360 + 1{\cdot}96 \times 12$ *i.e.* between 336 and 384. This shows that the investigator is quite likely to get between 336 and 384 people *i.e.* 0·56 and 0·64, approving the motion, although in 5 per cent of samples he will get a value outside these limits.

This latter example, while indicating the type of variation that might be expected in sampling to find a proportion, is the reverse of what is usually experienced in practice. The more usual occurrence is that the proportion of approvals in the sample is known, but it is not known how accurately this gives the proportion in the population. This problem is considered in the next section.

2A.8 *Accuracy of an estimated proportion*—Suppose, in a sample of 600 people, 360 approve a certain motion and it is required to find the limits within which the true proportion in the population lies. If the true proportion were 0·64, then in the manner of the last section, the expected number of approvals in the sample is 384 and the standard deviation is $\sqrt{(600 \times 0{\cdot}64 \times 0{\cdot}36)} = 11{\cdot}8$ so that, by *Table I*, 97·5 per cent of the samples would exceed $384 - 1{\cdot}96 \times 11{\cdot}8$ *i.e.* 360·9. Thus, if the true proportion exceeds 0·64, to get only 360 approvals is a rare event (occurring in less than 2·5 per cent of samples) and we can say with a certain degree of confidence that the true proportion is less than 0·64. As in section 2.5, if we conclude that the true proportion is less than 0·64, the possibility of this proportion exceeding 0·64 is not completely ruled out, but it is an unlikely event. Similarly, if the true proportion were 0·56, in 97·5 per cent of samples less than 359·8 approvals would be obtained. So that we know with a reasonable degree of certainty that the true proportion lies between 0·56 and 0·64. The degree of certainty is usually measured by the fact that this sample will occur in less than 2·5 per cent of trials when the true proportion exceeds 0·64, and in less than 2·5 per cent of trials when the true proportion is less than 0·56. The true proportion is in consequence said to lie between 0·56 and 0·64 with 95 per cent $(= 100 - 2{\cdot}5 - 2{\cdot}5)$ certainty.

This result might be compared with the result given in the last section, where it may be seen that the true proportion stands in almost the same relation to the sample proportion as the sample proportion does to the true proportion. The relationship is not completely symmetrical since the true proportion should always be used in finding the standard deviation, but the estimated proportion may be used if it is a reasonably accurate estimate of the true proportion *i.e.* if the sample is large. This approach is of importance in its practical applications as demonstrated by the following examples.

a Suppose 430 out of 1,000 people state a preference for a particular political party. Then the standard deviation of the number is $\sqrt{(1{,}000 \times 0{\cdot}43 \times 0{\cdot}57)} = 15{\cdot}7$, so that the proportion preferring this party may be stated as $0{\cdot}430 \pm 0{\cdot}0157$. This means that we are 95 per cent certain that the true proportion is between $0{\cdot}430 - 1{\cdot}96 \times 0{\cdot}0157$ and $0{\cdot}430 + 1{\cdot}96 \times 0{\cdot}0157$ *i.e.* 0·399 and 0·461, and 99 per cent certain that it is between $0{\cdot}430 - 2{\cdot}58 \times 0{\cdot}0157$ and $0{\cdot}430 + 2{\cdot}58 \times 0{\cdot}0157$ *i.e.* between 0·389 and 0·471.

b Many tests of the quality of commercial processes are destructive, as, for example, in testing shells, electric light bulbs, tyres or food, so that a sample of the goods has to be used in such tests.

Suppose in a batch of 300 articles, 12 articles are defective, then the standard deviation of the number of defectives is $\sqrt{(300 \times 0{\cdot}04 \times 0{\cdot}96)} = 3{\cdot}4$ and the proportion lies between $(12 - 1{\cdot}96 \times 3{\cdot}4)/300$ and $(12 + 1{\cdot}96 \times 3{\cdot}4)/300$ *i.e.* between 0·017 and 0·063, with 95 per cent certainty. These are fairly wide limits and in order to improve the accuracy of the estimate more articles would have to be tested. In this manner it is possible to control the quality of articles produced and to detect any deterioration in the production process.

c In seed mixtures there are usually some 'hard' seeds which will not germinate with the other seeds. The proportion of these seeds may be determined by using a sample of the seeds, and the accuracy of this determination may be found at the same time. For example, suppose out of some 500 seeds 50 fail to germinate, then the standard deviation of the number of hard seeds is $\sqrt{(500 \times 0{\cdot}10 \times 0{\cdot}90)} = 6{\cdot}7$, and the proportion of hard seeds is $0{\cdot}10 \pm 0{\cdot}0134$ *i.e.* we are 95 per cent certain that it is between 0·074 and 0·126.

2A.9 *Theoretical distributions: the Poisson*—A particular case of the binomial distribution occurs when the probability of a success is very small. In order to have any successes it is necessary to have a large number of trials, and provided the mean number of successes is large, the distribution tends to normality. However, if the mean number of successes is small a different distribution arises. This is the Poisson distribution: if the mean number of successes in a series of trials is m, then for the Poisson distribution the probability of getting r successes in a series of trials is $m^r e^{-m}/r!$

The Poisson distribution was first encountered as the distribution of the numbers of accidents happening to a large group of people during a fixed period. The probability of any particular person having an accident during the period was small but, since the number of people was large, some accidents usually occurred, and the numbers conformed to the Poisson distribution. However, other applications of this distribution are frequent. In bacteriological work, the distribution of the numbers of bacterial colonies observed on a plate usually conforms to the Poisson distribution, since the number of colonies that might be included in the sample is large but the probability of including any particular colony is small. Similarly, this distribution can be applied to the numbers of particles hitting a Geiger-Müller counter, to insect counts or to counts of diseased plants. In these examples, r becomes the mean number of particles hitting the Geiger-Müller counter in unit time, the mean number of insects observed in each count or the mean number of plants per unit area, and the probability of r particles hitting the counter in unit time, r insects being counted, or r diseased plants being observed per unit area, is $m^r e^{-m}/r!$ *Figures 16* and *17* give two distributions observed in practice, and the theoretical

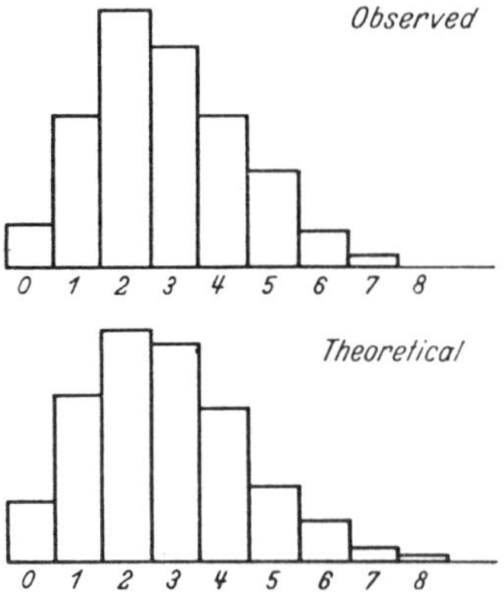

Figure 16. Frequency distribution of the number of bacterial colonies observed in 240 microscopic fields

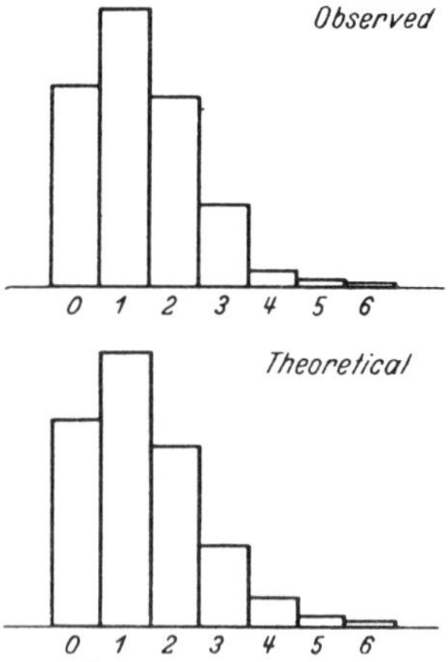

Figure 17. Frequency distribution of yeast cell counts made with a haemacytometer [Data of STUDENT, E. *Biometrika* 5 (1907) 351]

Poisson distributions which are most similar to these distributions.

The Poisson distribution when it arises requires special attention since it is determined once its mean is known. The appropriate methods of analysis will be discussed in Chapter 8.

SUMMARY OF PP 30 TO 36

The distribution of the number of successes in a series of trials has been considered. It has been shown that if p is the probability of success in a single trial, then the number of successes in n trials is distributed approximately normally with mean np and standard deviation $\sqrt{[np(1-p)]}$. Using this fact, it has been seen how the accuracy of an estimated proportion can be determined and used in sampling inquiries.

A second theoretical distribution, the Poisson, has been shown to arise in practice under certain conditions.

EXAMPLES

16 270 people out of 576 held a certain opinion. Show that this is not inconsistent with a majority of the population holding this opinion and that if there were a slight majority holding the opinion as low a value as 270 would be observed in 6·7 per cent of samples.

17 If 51 per cent of babies born are male and about 90,000 babies are born in Scotland every year show that the proportion of male births lies between 0·5057 and 0·5143 with 99 per cent certainty.

18 In order to estimate the proportion of deformed trees in a forest area, 1,000 trees are observed of which 56 are deformed. Show that the proportion is $0{\cdot}056 \pm 0{\cdot}0073$ and use this to conclude that the true proportion lies between 0·0417 and 0·0703 with 95 per cent certainty.

19 In a sample of 130 students, 29 had an intelligence quotient exceeding 120; show that the percentage of students with intelligence quotient over 120 is $22{\cdot}30 \pm 3{\cdot}65$ per cent.

20 Using the data of example *10* show that $9{\cdot}52 \pm 0{\cdot}64$ per cent of unmarried women of this city receive incomes of less than £100.

2A.10 *A rough estimate of the standard deviation*—Table I of the Appendix shows that, for a normal distribution, 6·68 per cent of observations exceed a normal deviate of 1·5, and a further 6·68 per cent fall below a normal deviate of −1·5. Thus, for a normal distribution, $2 \times 6{\cdot}68$ per cent $= 13{\cdot}36$ per cent or, roughly one on eight, of all observations has a normal deviate exceeding 1·5 in absolute value. To put this another way, the central seven-eighths of the observations have a range of roughly $2 \times 1{\cdot}5 = 3$ times the standard deviation.

To gauge the standard deviation of a series of observations, one may thus find the value a exceeded by roughly one-sixteenth of the observations, and the value b below which one-sixteenth of the observations fall, then $\frac{1}{3}(a-b)$ gives a rough estimate of the standard deviation.

This rough rule has two important properties. First, with a large number of normally-distributed observations, it gives rise to the best method of estimating a standard deviation employing a simple range. Secondly, the method tends to hold good even when the observations are not exactly normally distributed.

Consider, for example, the 225 observations shown in Figure 3. Now, one-sixteenth of 225 is roughly 14, and, in Figure 3, 17 observations exceed the value 132·5 and 10 fall below 87·5. Thus the standard deviation is roughly $\frac{1}{3}(132{\cdot}5-87{\cdot}5) = \frac{1}{3}(45) = 15$* (cf. 14·2).

*In fact, interpolation might be used to get slightly more precise estimates of the values of a and b, in which case we get $a = 134$ and $b = 90$ and the estimated standard deviation is now $\frac{1}{3}(44) = 14{\cdot}7$.

Alternatively, for example 1, one-sixteenth of 460 is roughly 29, and there are 24 trees with a girth below 29 in. and 36 trees with a girth above 61 in. Thus the standard deviation is roughly $\frac{1}{3}(61-29) = \frac{1}{3}(32) = 10{\cdot}7$ (cf. 10·8).

Application of this rule may be used to check for gross errors in the full calculation of the standard deviation, or in some circumstances in the first stages of a statistical investigation to gauge whether more extensive calculations might be warranted.

3

COMPARISON OF TWO SETS OF MEASUREMENTS

3.1 *Method of comparison*—It has been shown that the mean and standard deviation provide a convenient summary of a set of measurements and, consequently, that a comparison between two sets of measurements can most easily be made by using their estimated means and standard deviations. If the number of observations in each group is sufficiently large then the estimates are accurate and any difference between the estimated means or standard deviations is real, but when the number of observations is not large then a difference may be due to chance. For example, one group of ten children might have a mean weight of 110 lb while a similar group, by pure chance, has a mean weight of 120 lb (in the same manner as one child may have a weight of 100 lb and a second child a weight of 130 lb). Similarly, natural variability will complicate comparison of the estimates of standard deviations. It is therefore necessary to allow for chance effects in comparing sets of observations to ensure that differences are real.

Usually we wish to test the difference between two means, but this is dependent upon the variability or standard deviation of the individual observations which must therefore be taken into account. Occasionally we wish to test the ratio of the means since any difference may tend to act proportionally, as for example with bacterial counts which are generally affected proportionally by any treatment. However, since on these occasions the logarithm of the measurement is usually employed, the difference between the logarithms (which is the logarithm of the ratio) is still an appropriate measure.

The comparison of two standard deviations is much simpler, since it is the ratio of two spreads that concerns us and this is a dimensionless quantity dependent only upon the numbers of observations in each set. A comparison of two means is more frequently required, but since the comparison of standard deviations is a simpler test this will be considered first.

3.2 *Variance-ratio test*—If each standard deviation was determined from a very large number of observations then any deviation of their ratio from unity would be a real deviation. However, since both standard deviations are usually subject to variability, their ratio is also subject to variability.

The extent of this variability can be predicted from a knowledge of the normal distribution. Thus, for example, if two samples of ten observations are drawn from the same population and the standard deviation estimated for each sample in a series of trials, the proportion of trials in which the ratio of the standard deviations exceeds any value can be found as in *Table 3.1*.

Table 3.1. Distribution of the Ratio of Two Estimated Standard Deviations

Value	0·31	0·39	0·43	0·50	0·56	0·64	0·79	1·00
Percentage of trials in which value is exceeded	99·9	99·5	99·0	97·5	95·0	90·0	75·0	50·0
Value	1·26	1·56	1·79	2·01	2·31	2·56	3·19	
Percentage of trials in which value is exceeded	25·0	10·0	5·0	2·5	1·0	0·5	0·1	

Thus 50 per cent of the ratios will exceed 1·00, and 10 per cent will exceed 1·56 by pure chance. If therefore we get a ratio of, say, 2·6, we have either to conclude that there is a real difference between the two samples or that an unlikely event (which would occur about once in two hundred times by pure chance) has happened and that there is no difference between the samples. Usually we will conclude the former, and say that the difference is 99·5 per cent significant or, alternatively, significant at the 0·5 per cent level *i.e.* it would occur by pure chance in 0·5 per cent of trials. However, if the ratio is, say, 1·5, we have to conclude either that there is a real difference between the two samples or that an event which would occur about once in ten times by pure chance has happened and that there is no difference between the samples. Here, both conclusions are quite probable, so that judgement is deferred and we say that the difference between the standard deviations is not significant *i.e.* it may be due to chance.

The values in *Table 3.1* vary according to the numbers of observations used to determine each standard deviation, and a large table has been constructed to give these values for each combination of these numbers. Various modifications are used in the presentation of this table. First, instead of dealing with the ratio of the standard deviations, the ratio of the variances is used. This ratio has the advantage that in the calculation it is not necessary to take the square roots. Secondly, since no estimate of the variance can be made from one observation the effective number of observations is used in entering the table. This is one less than the number of observations and is usually called the number of degrees of freedom*. Lastly, the values exceeded by pure chance in, say, 5 per cent, or any other percentage of trials, are collected together in one table.

* This is used to divide $\Sigma (x-\bar{x})^2$ in order to estimate the variance.

Thus a set of tables, such as *Tables III* and *IV*, gives the required significance levels.

These tables should be used when we wish to test whether a first standard deviation is greater than a second, but not *vice versa.* It will be seen later that this is the more common occurrence. If, however, we wish to test whether a first standard deviation differs significantly from a second it is necessary to take into account the possibilities that the first may be larger or smaller than the second. The percentage given by the table must then be doubled.

3·3 *Examples of use of variance-ratio test*—Example 5 gave the standard deviations of four sets of ten indices of pig-iron production, the values being 5·46, 24·74, 17·60, and 5·95. These might be compared using *Table 3.1*. Since we are interested in both increases and decreases, we shall conventionally divide the larger standard deviation by the smaller and use twice the percentage given by the table. The ratios of successive pairs of standard deviations are 4·53, 1·41, and 2·92, and from *Table 3.1* these would occur by pure chance in roughly 0·2, 50, and 0·5 per cent of trials. Thus the first and last ratios would be judged significant but not the second.

If the test is now carried out using the variance-ratio tables, the variances are 29·77, 611·91, 309·99 and 35·43, and the ratios become 20·55, 1·97, and 8·75, the denominators and numerators of which both have nine degrees of freedom. From *Tables III* and *IV*, the value 3·18 would be exceeded in 10 per cent and 5·55 in 2 per cent of trials; the percentages being doubled as previously. It is quite clear that the first and third ratios are significant but not the second. In order to determine the significance more accurately a more extensive set of tables would have to be used, but for most practical purposes the 5 per cent and 1 per cent tables will suffice.

As a second illustration, the variabilities in the girths given in example 1 might be compared. The standard deviations were 10·78 and 9·75 in and the variance ratio is thus $(10{\cdot}78/9{\cdot}75)^2 = 1{\cdot}22$ which now has 459 and 23 degrees of freedom for its numerator and denominator. Referring to *Table III* with 459 and 23 degrees of freedom we see that 1·76 is exceeded by pure chance in 10 per cent of trials. We have fallen very short of this value, so that the value 1·2 can be ascribed to chance and we conclude that the variabilities in the girths of the two groups are not significantly different.

Lastly, consider the variance ratio for the two sets of observations in example 2. This is $(22{\cdot}5/20{\cdot}3)^2 = 1{\cdot}23$ and both numerator and denominator have 99 degrees of freedom. Since from *Table III*, the value 1·39 is exceeded in 10 per cent of cases, this ratio would occur too frequently by pure chance to allow us to conclude that there is a real difference between the two groups.

3.4 *Pooling of variances*—If it is decided that the estimates of variance in several sets of observations do not differ significantly, it is usually desirable to combine them to give a better estimate. This is most advantageous when the numbers of observations in each set are small and the individual estimates of variance are inaccurate as a result. The means of each set of observations need not coincide for a combined estimate of variance to be obtained. In fact, this is the more usual state of affairs.

The variance for a single set of observations is found from the sum of squares of deviations of each observation about its mean divided by the effective number of observations *i.e.* the degrees of freedom. Similarly, the variance for several groups of observations is found from the sum of squares of deviations of observations from each group mean divided by the effective number of observations *i.e.* the total of the degrees of freedom for each group. For example, the two sets of observations, 11, 9, 10, 9, 8; and 15, 18, 16, 16, 17, 14 deviate from 10 by 1, −1, 0, −1, −2 and 5, 8, 6, 6, 7, 4, so that their variances are

$$\tfrac{1}{4}(1+1+0+1+4-3^2/5)=5{\cdot}2/4=1{\cdot}30$$

and

$$\tfrac{1}{5}(25+64+36+36+49+16-36^2/6)=10/5=2{\cdot}00$$

with 4 and 5 degrees of freedom respectively. The variance ratio $2{\cdot}00/1{\cdot}30 = 1{\cdot}54$ falls far short of the value 6·26 which is exceeded by pure chance in 10 per cent of trials. The two variances may thus be combined to give a more accurate estimate with 9 degrees of freedom

$$\frac{10{\cdot}0+5{\cdot}2}{5+4}=\frac{15{\cdot}2}{9}=1{\cdot}69$$

If a third set of observations 6, 5, 2, 7 is taken, then the estimated variance from these is

$$\tfrac{1}{3}(36+25+4+49-20^2/4)=14/3=4{\cdot}67$$

Although this is nearly three times the above estimate it is not significantly greater at the 10 per cent level, 3·86. We may thus use an overall estimate $(14+15{\cdot}2)/(3+9)=2{\cdot}43$ with 12 degrees of freedom.

Mathematically, if the first set of n_1 observations is $x_1, x_2 \ldots$, the second set of n_2 observations is $y_1, y_2, \ldots$, then the combined variance is given by

$$s^2=\frac{\Sigma(x-\bar{x})^2+\Sigma(y-\bar{y})^2+\ldots\ldots\ldots}{n_1-1+n_2-1+\ldots\ldots\ldots}$$

3·5 *Accuracy of arithmetic mean*—Before comparing arithmetic means it is necessary to investigate the reliability of the arithmetic mean and the extent to which it is likely to vary. This, in itself, is important since it is very necessary to know how accurate an estimate of the mean is likely to be, and how many measurements have to be taken to achieve a particular degree of accuracy. For example, the forester using a sample to estimate the volume of timber in a stand of trees will want to know if sufficient trees have been measured to ensure an accurate result, and if not, how many trees must be measured. Likewise, in making measurements of biological, physical and economic constants the reliability of the final estimates is often as important as the estimated constants.

If a set of observations is repeated a different value of the arithmetic mean will usually be obtained. If it is repeated a sufficient number of times a distribution of the arithmetic means will be obtained. Further, if each observation is determined by a large number of small effects so that the distribution of the individual observations is normal, then the arithmetic mean is likewise determined by a large number of small effects, and is distributed normally as a consequence. In fact, even if the individual observations are not distributed normally the mean will tend to be distributed normally, since each observation contributes in part to the arithmetic mean which is consequently determined by many more effects than the individual observations.

Figure 18 gives the observed distributions of individual tree girths, mean girths of groups of four trees and mean girths of groups of sixteen trees. It is obvious from this figure that the means of four trees and of sixteen trees are distributed normally with standard deviations equal to one half and one quarter of the standard deviation of the individual tree girths.

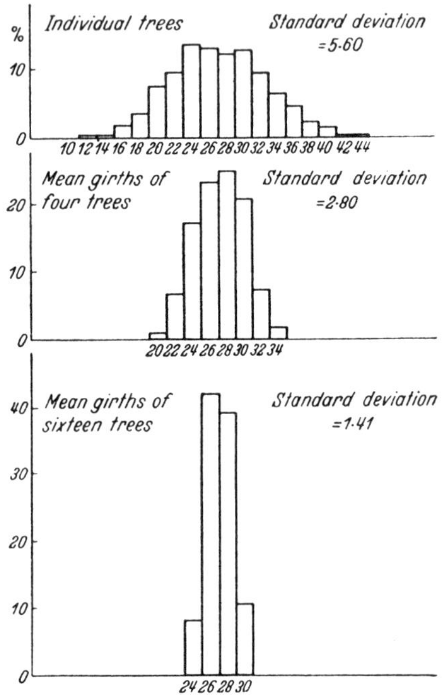

Figure 18. Frequency distributions of individual tree girths, mean girths of four trees and mean girths of sixteen trees

This observation is important since it indicates that the taking of four times as many observations gives only twice as accurate an estimation of the mean. This is further illustrated in *Figure 19*, which gives the distributions of the means on twice and four times the original scale.

It can be shown that this illustration is a special case of a theorem which states:

The variance of the sum or difference of a series of independent measurements is the sum of their variances.

This theorem is proved in section 3A.13. From it we see that the sum of a set of n observations, each with a variance σ^2, has a variance of $n\sigma^2$ or standard deviation $\sigma\sqrt{n}$. Consequently the arithmetic mean of a set of n independent observations has a standard deviation $(\sigma\sqrt{n})/n=\sigma/\sqrt{n}$ It is this fact which is demonstrated for $n=4$ and 16 in *Figures 18* and *19*. Using this theorem we are therefore able to calculate the accuracy of the arithmetic mean of a set of observations from the standard deviation of the observations.

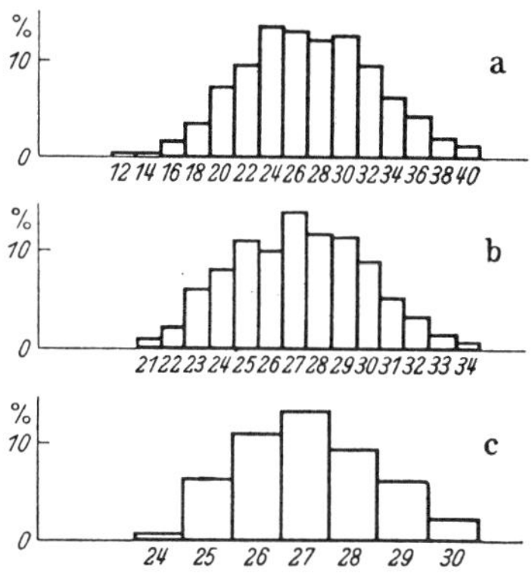

Figure 19. Frequency distributions of individual tree girths, mean girths of four trees and mean girths of sixteen trees

In order to avoid possible confusion between the terms standard deviation and standard deviation of the mean, the former term is applied exclusively to the original observations, while the latter is usually called the standard error of the mean, or, briefly, the standard error.

3.6 *Examples of determination of accuracy of arithmetic mean*—In the following examples, sufficient observations *i.e.* more than fifty, have been taken to provide a fairly accurate estimate of the standard deviation. If this is not so, then the possible inaccuracy of the estimate must be taken into account by a correction which is described in section 3.9.

a The arithmetic mean and standard deviation of the scores of 100 male students, given in example *2*, were 0·5 and 20·3, and consequently the standard error of the mean is $20{\cdot}3/\sqrt{100}=2{\cdot}03$. Now, from *Table II*, it is known that in 95 per cent of trials the normal deviate will be less than 1·96 or, alternatively, that the true mean will deviate from the observed mean by not more than $1{\cdot}96\times2{\cdot}03=4{\cdot}0$ Thus we are 95 per cent certain that the mean lies between 0·5 $-$4·0 and 0·5 + 4·0 *i.e.* between $-4{\cdot}5$ and 3·5. Similarly, we are 99 per cent certain that the true mean lies between $0{\cdot}5-2{\cdot}58\times2{\cdot}03$ and $0{\cdot}5+2{\cdot}58\times2{\cdot}03$ *i.e.* between $-5{\cdot}7$ and 4·7, and, corresponding to any degree of certainty, limits can be set for the true mean.

b The weekly expenditure on rent was estimated for a random sample of 80 middle class families as 29 *s* with a standard deviation 6 *s*. The standard error of this estimate is $6/\sqrt{80}=0{\cdot}67$ *s* and we are 95 per cent certain that the true mean weekly expenditure lies not more than $1{\cdot}96\times0{\cdot}67=1{\cdot}31$ *s* from the estimated mean of 29 *s*. Hence we are 95 per cent certain that the true mean lies between 27·69 and 30·31 *s*, and in the same manner we are 99·99 per cent certain that it lies between $29-3{\cdot}89\times0{\cdot}67$ and $29+3{\cdot}89\times0{\cdot}67$ *s* *i.e.* between 26·39 and 31·61 *s*.

The determination of the accuracy of the arithmetic mean in the manner used in these two examples allows us to set limits within which the true mean of a population falls with any degree of certainty. The more certain we require to be, the wider will be the limits. Thus, for 95 per cent certainty, roughly twice the standard error is marked off on either side of the estimated mean; for 99·7 per cent certainty, roughly three times the standard error is used; and for 99·994 per cent certainty, roughly four times the standard error is required. Limits of this kind may be termed fiducial or confidence limits*. To specify the degree of certainty the percentage may be indicated and we may speak of 95 per cent fiducial or confidence limits.

3·7 *Comparison of arithmetic means*—The difference between two arithmetic means is subject to variability in the same manner as the individual observations or means, and we have to decide whether any difference is a chance difference or a real effect. In this, we are helped by the knowledge that the difference will be normally distributed, since the effects contributing to the normality of the means will also contribute to the normality of their difference: also by the theorem given in section 3·5, its variance will equal the sum of the variances of the two means. Hence, if σ_1 and σ_2 are the standard deviations of two groups of n_1 and n_2 observations, the standard error of the difference between their arithmetic means is

$$\sqrt{\left(\frac{\sigma_1^2}{n_1}+\frac{\sigma_2^2}{n_2}\right)}$$

When the standard deviations of the two sets of observations are the same this becomes

$$\sigma\sqrt{\left(\frac{1}{n_1}+\frac{1}{n_2}\right)}$$

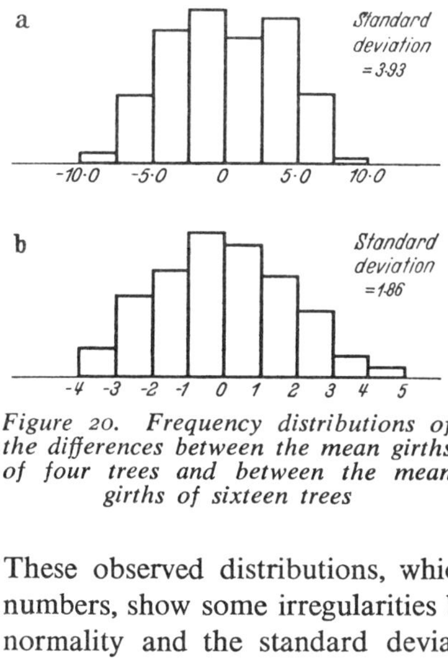

Figure 20. Frequency distributions of the differences between the mean girths of four trees and between the mean girths of sixteen trees

Figure 20 gives observed distributions of the differences between the means of the girths of four trees and the means of the girths of sixteen trees. These observed distributions, which are based upon comparatively small numbers, show some irregularities but they do not deviate appreciably from normality and the standard deviations of these distributions agree very well with the expected values $5{\cdot}60\ \sqrt{(\frac{1}{4}+\frac{1}{4})}=3{\cdot}96$ and $5{\cdot}60\ \sqrt{(\frac{1}{16}+\frac{1}{16})}=1{\cdot}98$.

* There is, in fact, a subtle difference between fiducial and confidence limits, and these are, strictly speaking, fiducial limits, although the confidence limits take the same values. In this book, fiducial and confidence limits will always coincide. This, however, is not always true and in some more advanced applications of statistical methods the two sets of limits may be distinguished.

To test the difference between means we assume that the means are from the same population so that the distribution of the difference has a zero mean. Using this distribution we calculate the probability of getting as large a difference as that observed. If this probability is small we reject the null hypothesis that there is no difference between the means. Alternatively, we can, in the manner of the last section, calculate the limits within which the difference may vary and thus ensure that it could not be zero. Both of these methods are considered below.

In using these methods to test the difference between two treatments, methods or factors we must naturally ensure that no other factors enter into the comparison, and this is done by assigning the treatments or methods at random. Thus in an agricultural field trial the treatments or varieties are assigned to the plots at random to ensure that no large fertility differences can enter into the final comparison and that the differences which do occur are taken into account in estimating the variance. Likewise two series of observations should be taken in random order if they are to be compared, so that any change in conditions half way through the observations will not affect their comparison.

By random order it is meant that the two series of observations conform to no pattern, as opposed to systematic or ordered sets of observations. Hence to take the observations for each set alternately would not be random; the comparison of the two sets may be violated if alternate observations coincide with day and night, or with some unsuspected periodic factor. This disadvantage outweighs the advantage that systematic sets of observations cover the area, material or period of time sampled uniformly. It will be seen in subsequent chapters that we may achieve this uniformity by appropriate experimental designs and analysis and still conserve a random element in the method of sampling.

3.8 *Examples of test of difference between means*—In demonstrating applications of the above theory, examples have again been chosen with sufficient observations to give a reliable estimate of the standard deviation. The correction to be applied when the number of observations is small and, as a consequence, the standard deviation is not accurately estimated is considered in the next section.

a The means and variances of the scores obtained by students in a psychological test for assertiveness, mentioned in example *2*, were 0·5 and 411 for the males and 10·7 and 506 for the females. The difference between the means is 10·2 and its standard error is $\sqrt{\left(\frac{411}{100}+\frac{506}{100}\right)}=3{\cdot}0$.

On the assumption that the true difference is zero, the normal deviate is $10{\cdot}2/3{\cdot}0=3{\cdot}4$, and since, from *Table II*, as large a value as 3·29 would occur by pure chance in less than 0·1 per cent of trials, the deviate of 3·4 is unlikely to have arisen by pure chance. We therefore reject the null hypothesis and conclude that the difference between the means is real.

Alternatively, in the manner of section 3.6, we can set limits for the true value of the difference. For example, we are 99·9 per cent certain that the true difference lies between $10{\cdot}2-3{\cdot}29\times3{\cdot}0$ and $10{\cdot}2+3{\cdot}29\times3{\cdot}0$ *i.e.* between 0·3 and 20·1, so that we can say with this degree of certainty that the difference is not zero.

b The mean weight increases during a 40-day period of two groups of 30 male rats on cannibal and vegetarian diets were 161·2 and 154·1 gm, and the variances of the weight increases were 341·9 and 412·3 gm.

The variance ratio of 1·21 might easily be due to chance so a combined variance of $(341{\cdot}9+412{\cdot}3)/2=377{\cdot}1$ with 58 degrees of freedom can be used. However, since the numbers in the two groups are equal, the pooling of the variances does not affect the final result. The standard error of the difference between means is then $\sqrt{[377{\cdot}1\,(1/30+1/30)]}=5{\cdot}0$, and the normal deviate corresponding to the difference of 7·1 gm is $7{\cdot}1/5{\cdot}0=1{\cdot}42$. From *Table II* the deviate 1·44 is exceeded in 15 per cent of trials so that this difference might easily be due to chance.

3.9 *The t test*—It has been pointed out that when the numbers of observations are small, the variances will be inaccurately determined and in testing the accuracy of a mean or the difference between means allowance will have to be made for this fact. If the variances of the two samples are different, both estimates may be inaccurate, and the correction to be made depends upon their degrees of freedom and also upon the ratio of the variances; but, if the variances are equal, the correction to be made depends only upon the degrees of freedom of the pooled variance. Only the latter problem, which is the simpler and more common, will be considered here. A discussion of the appropriate methods when the variances of the two samples are different will be postponed until Chapter 8.

If the joint variance of the two samples is determined exactly *i.e.* the number of degrees of freedom is effectively infinite, then using *Table II* we can express the percentage of trials in which a particular deviate is exceeded (ignoring the sign), in an abbreviated form similar to *Table 3.1*:

Table 3.2

Percentage of trials in which deviate is exceeded	50	25	10	5	2·5	1·0	0·5	0·1
Deviate	0·67	1·15	1·64	1·96	2·24	2·58	2·81	3·29

When the variance is based upon a small number of degrees of freedom these values have to be revised. For example, if the variance is based upon 50 degrees of freedom the deviate exceeded in 5 per cent of trials is 2·01 instead of 1·96. This rises to 2·09 for 20 degrees of freedom, 2·23 for 10 degrees of freedom, 2·57 for 5 degrees of freedom, and 12·71 for a single degree of freedom. Thus if the variance is estimated with less accuracy the limits assigned to the true mean have to be widened. These values of the normal deviate when the estimated variance is used are called *t* values. In the same way *Table V*, which gives the *t* values corresponding to given percentages and degrees of freedom, is called a *t* table.

The following examples demonstrate the use of this table.

a In a wheat variety trial, two varieties were allotted to eight 1/40-acre plots at random. The yields of grain from these plots were 17·1, 21·2, 19·7 and 18·4 lb for variety *A*, and 18·1, 14·3, 16·7 and 16·1 lb for variety *B*.

The mean yields for these varieties are 19·1 and 16·3 lb, so that the variances are $[2{\cdot}0^2+2{\cdot}1^2+0{\cdot}6^2+0{\cdot}7^2]/3=9{\cdot}26/3=3{\cdot}09$ and $[1{\cdot}8+2{\cdot}0^2+0{\cdot}4^2+0{\cdot}2^2]/3=7{\cdot}44/3=2{\cdot}48$. These are not significantly different (as might be expected, since the same factors will affect the plot yields of each variety) so that we can obtain a pooled estimate of variance $(9{\cdot}26+7{\cdot}44)/6=2{\cdot}78$ with six degrees of freedom. The standard error of the difference between the means is thus $\sqrt{[2{\cdot}78\,(\frac{1}{4}+\frac{1}{4})]}=1{\cdot}36$ and the estimated deviate t is $(19{\cdot}1-16{\cdot}3)/1{\cdot}36=2{\cdot}06$.

Referring to *Table V* with 6 degrees of freedom, we see that as high a value as 1·94 occurs in 10 per cent of trials, and 2·45 in 5 per cent of trials (compared with the values 1·64 and 1·96 when the variance is accurately determined). Thus a value exceeding 2·06 occurs by pure chance in about 9 per cent of trials and we cannot conclude that the difference between the varieties is significant.

b In example 5, the mean indices of pig-iron production for the first and last periods are 105·0 and 102·8, and the pooled variance for these periods is $(29{\cdot}77+35{\cdot}43)/2=32{\cdot}60$ with $9+9=18$ degrees of freedom. The standard error of the difference between the means is $\sqrt{[32{\cdot}60\,(1/10+1/10)]}=2{\cdot}55$, and the value of t is $(105{\cdot}0-102{\cdot}8)/2{\cdot}55=0{\cdot}86$. From *Table V*, as high a value as this would occur by chance in 40 per cent of trials, so that the mean levels of production in the two periods are comparable.

c The percentages of clay were observed for 12 freely drained and 10 poorly drained basal horizons. These were found to be 8·1, 11·3, 15·1, 12·1, 18·2, 5·1, 14·3, 9·2, 9·6, 13·8, 7·9, 12·8, for the freely drained soil and 17·8, 15·3, 7·9, 18·7, 15·1, 22·3, 9·5, 17·8, 16·6, 14·1, for the poorly drained soil. The totals and means of these two sets of values are 137·5 and 11·46 for the former, and 155·1 and 15·51 for the latter. The estimated variance of the clay percentages in the freely drained soil is

$$\frac{1}{11}\left[8{\cdot}1^2+11{\cdot}3^2+\ldots+12{\cdot}8^2-\frac{137{\cdot}5^2}{12}\right]=\frac{147{\cdot}43}{11}=13{\cdot}40$$

while on the poorly drained soil it is

$$\frac{1}{9}\left[17{\cdot}8^2+15{\cdot}3^2+\ldots+14{\cdot}1^2-\frac{155{\cdot}1^2}{10}\right]=\frac{164{\cdot}19}{9}=18{\cdot}24$$

The variance ratio, $18{\cdot}24/13{\cdot}40=1{\cdot}36$, with 9 and 11 degrees of freedom is not nearly significant, so that a pooled variance, $(147{\cdot}43+164{\cdot}19)/(11+9)=15{\cdot}58$ with 20 degrees of freedom, can be used. The standard error of the difference between the means is thus $\sqrt{[15{\cdot}58\,(1/12+1/10)]}=1{\cdot}69$, and the estimated deviate t is $(15{\cdot}51-11{\cdot}46)/1{\cdot}69=2{\cdot}40$. Entering *Table V* with 20 degrees of freedom, we see that as high a value as 2·42 occurs by chance in 2·5 per cent of trials. In consequence, as large a difference as has been observed would occur by pure chance roughly once in forty times and we conclude that the difference is likely to be real.

In practice we may often take for granted the equality of the variances for the two groups. Where the measurements are similar and affected by the same extraneous variation we do not always test the equality of the variances. It is, however, always advisable to maintain at least a rough check on this assumed equality.

SUMMARY OF PP 38 TO 47

It has been shown that variances can be compared using a variance-ratio table, from which the size of the ratios likely to arise by pure chance can be judged.

The accuracy of the mean of n observations can be determined from the fact that it is normally distributed with standard error $\sigma/\sqrt{n}$. The comparison of the means of two sets of n_1 and n_2 observations with standard deviations σ_1 and σ_2 can be made using the fact that the difference between the means is normally distributed with standard error

$$\sqrt{\left(\frac{\sigma_1^2}{n_1}+\frac{\sigma_2^2}{n_2}\right)}$$

When the numbers of observations are small so that the variances are not accurately estimated, then this has to be taken into account. Most commonly, when the variances of the two sets of observations are the same *i.e.* $\sigma_1=\sigma_2$, this is done by using a t table instead of a table of the normal deviate.

EXAMPLES

21 The weight increases of eight male rats from the same litter during a month were 127·0, 112·8, 114·1, 123·1, 119·2, 116·4, 116·2 and 121·2 gm. A second group of male rats from different litters increased in weight by 115·6, 119·7, 133·1, 102·3, 127·6, 129·3, 107·8 and 111·3 gm. Show that the variances of the weight increases in these two groups are 23·18 and 121·57, and use these to demonstrate that an experiment on weight increases using rats from the same litter is likely to be more accurate.

22 The mean and standard deviation of the weights of a sample of 68 18-year old Aberdeen students in 1947 were 146·66 and 12·79 lb. Show that if all of the students of this age had been weighed, their mean weight would lie between 143·56 and 149·76 lb with 95 per cent certainty.

23 The wireworm population of a field is normally estimated by counting the number of wireworms in 4-inch cylindrical cores taken at random from a field from which the total wireworm population of the field can be estimated; 50 such cores taken from a field had a mean count of 3·52 and a standard deviation of 3·15. Show that the mean has a standard error of 0·445, and conclude that the true population can be determined to within 33 per cent with 95 per cent certainty.

24 Using the data of example *9*, show that the difference of 0·09 years between the mean ages of Bristol mothers at *primiparae* in 1932 and 1937 has a standard error of $\left[\frac{(4\cdot63)^2}{2{,}051}+\frac{(4\cdot75)^2}{2{,}517}\right]=0\cdot14$ years, and hence conclude that the difference could be ascribed to natural variability. Show also that a difference of 4 months would have been regarded as significant.

25 Six independent estimates of the volume of timber on an estate gave values 1·14, 1·04, 1·13, 1·17, 1·23 and 1 19 million cu ft. Show that the mean and variance of these values are 1·15 and 0·0042 and conclude that the true value lies between 1·082 and 1·218 with 95 per cent certainty.

26 The bacterial counts (in millions) on herrings stored at two temperatures, 20° and 37°C, were estimated after 8 days. The counts on the herrings stored at 20°C were 7·9, 9·0, 6·9, 7·0, 6·5, 6·1 and 6·6, while the corresponding figures for those stored at 37°C were 7·6, 8·4, 9·8, 7·1, 6·3 and 7·8.

Show that the means and pooled variance of these counts are 7·14, 7·83 and 1·185, and that the standard error of the difference between these means is 0·606. Hence show that the estimated deviate t is 1·139 with 11 degrees of freedom, and conclude that this difference might be ascribed to chance.

EXTENDED DEVELOPMENT

3A.10 *Calculation of number of observations necessary for a given accuracy*—In assigning a standard error to an arithmetic mean or difference of means, we are determining the range within which it might vary. If we wish to narrow this range then it is necessary to take more observations and, to halve the range, four times the number of observations must be taken. Since it is often necessary to determine a mean with a certain degree of accuracy or to detect differences of a certain size, the number of observations necessary for this should be known, especially if the taking of observations is arduous or expensive. This we do by estimating the standard deviation from the first few observations and by equating the limits for an arbitrary number of observations to the required accuracy. The required number of observations may then be found. To demonstrate this method consider the following examples:

a It is required to estimate the mean girth of a group of trees to within 1 per cent. Suppose the first 100 trees measured have a mean girth of 27·1 in and a standard deviation of 5·6 in, then if we estimate the mean girth to within 0·27 in this is likely to be accurate enough. The standard error of the mean is $5{\cdot}6/\sqrt{n}$ and we are 95 per cent certain that for n observations the true mean lies within $1{\cdot}96 \times 5{\cdot}6/\sqrt{n}$ of the estimated mean. Thus, if $1{\cdot}96 \times 5{\cdot}6/\sqrt{n} = 0{\cdot}27$, or $n = (1{\cdot}96 \times 5{\cdot}6/0{\cdot}27)^2 = 1{,}640$, we are 95 per cent certain that the estimate is within 0·27 in of the true mean.

Alternatively, if $2{\cdot}58 \times 5{\cdot}6/\sqrt{n} = 0{\cdot}27$, or $n = (2{\cdot}58 \times 5{\cdot}6/0{\cdot}27)^2 = 2{,}840$, we are 99 per cent certain that the estimate is within 0·27 in of the true mean. These values will be affected slightly by the inaccuracy of the estimates but they give rough indications of the numbers of measurements required.

b In an experiment to determine the effect of grazing upon the percentage botanical composition of herbage, small plots, both grazed and ungrazed, are cut and compared. An initial analysis shows that the standard deviation of the measurements on each plot is about 3 per cent, and it is desired to detect differences of 1 per cent or greater in the percentage composition. Suppose n plots of each type are used, then the standard error of the difference is $3\sqrt{(1/n + 1/n)}$, and we are 95 per cent certain of detecting a 1 per cent difference if $1{\cdot}96 \times 3\sqrt{(2/n)} = 1$ *i.e.* $n = 2\,(1{\cdot}96 \times 3)^2 = 69$ plots of each kind must be used. Again this value should be regarded as indicating rather than determining the number of plots to be used.

3A.11 *Relative precision and combination of experimental results*—It frequently happens that several series of experiments or measurements are carried out under different conditions or by different methods so that each series of measurements has a different precision. When this occurs, it is useful to have an index of the relative accuracies or efficiencies of the various methods and such a measure is provided by the reciprocal or inverse variance. As a result of doubling the number of observations taken, the variance of the mean is halved. Thus, if the variance of one set of observations is double that of another, twice as many observations must be taken in the first set to obtain the same accuracy. On such occasions, we say that the efficiency of the second set is twice that of the first set and since the inverse variance of the second set is double that of the first set, this provides a comparison of the precisions of the observations. Alternatively, the inverse variance gives the equivalent number of observations with unit variance. Thus, for example, one observation with variance $\frac{1}{2}$ is equivalent to two observations with unit variance, and so are ten observations with variance 5, since each gives a mean with variance $\frac{1}{2}$.

If several sets of observations are to be combined, then in combining their means we will usually take into account the differing numbers of observations in each group. For example, suppose we have two groups of observations 2, 3, 4 and 4, 6 with means 3 and 5, the combined mean will not be $\frac{1}{2}(3+5)=4$ but $(2+3+4+4+6)/5=3{\cdot}8=(3\times 3+2\times 5)/(3+2)$ *i.e.* we 'weight' the means 3 and 5 in the ratio of the numbers of observations, taking three times the first estimate and twice the second estimate. In this manner we ensure that each observation is equally represented. However, it may happen that the sets of observations are of different accuracies due to changing conditions, to improvements in technique, or to changes in methods or observers. Under such conditions we shall not wish to represent the observations equally but to emphasize each according to its accuracy. This may be done by weighting each mean according to its inverse variance. To demonstrate how this works in practice, consider the following example:

Three sets of 6, 13 and 9 estimates of the physical quantity g by different methods had means 980·63, 981·27 and 980·84 and variances 1·27, 0·35, 0·82 respectively. The second method is more than three times as efficient as the first, and more than twice as efficient as the third method, and greater emphasis is consequently laid upon this when calculating the combined estimate as follows:

1 No. of observations	*2* Mean	*3* Variance	*4* Equiv. No. of obsns with unit variance = 1/3	*5* Total of equiv. obsns = 2 × 4
6	980·63	1·27	4·72	4,628·57
13	981·27	0·35	37·14	36,444·37
9	980·84	0·82	10·98	10,769·62
			52·84	51,842·56

The total of these 52·84 comparable observations of unit variance is 51,842·56 so that the combined mean is $51{,}842{\cdot}56/52{\cdot}84=981{\cdot}12$ and this has a variance of $1/52{\cdot}84=0{\cdot}0189$.

It should be noted that if the mean of the means

$$\frac{1}{3}(980{\cdot}63+981{\cdot}27+980{\cdot}84)=980{\cdot}91$$

had been used it would have had a variance

$$\frac{1}{3}\left(\frac{1{\cdot}27}{6}+\frac{0{\cdot}35}{13}+\frac{0{\cdot}82}{9}\right)=0{\cdot}1099$$

This estimate is therefore only one fifth as efficient as the above estimate. If the numbers of observations are taken into account, but not their differing accuracies, the efficiency is only three quarters of that of the above method.

3A.12 *Estimation of variance from observations of differing precision*— In combining several sets of observations in the last section, we used the inverse variance of each set of observations as an index of their relative accuracies. However it sometimes happens that we know the relative accuracies from the method of observation and therefore we do not need to use the variance as an index. Such a state of affairs occurs in agricultural experiments when the yield of several plots is bulked, since we know that the mean yield of the bulked plots will be proportionately more accurate than the individual plot yields. This is not to be generally recommended since the bulking of observations loses information on the variability and accuracy of the results, but sometimes the bulking cannot be avoided. Thus, sometimes in animal experiments since the herd or flock is the natural unit it is impossible to segregate the animals lest conditions become unreal. For example, we may be interested in the food intake of a sheep under pastoral conditions, but we may have to use flocks of sheep to preserve normal conditions. Ideally, each herd should be of the same size so that all comparisons between herds are of equal accuracy but this condition cannot always be realized in practice.

Thus it sometimes happens that the relative accuracies of observations are determined by the sizes of the samples taken and the best estimate of the mean is determined without use of the variance. Then, since the relative accuracies of the observations are known, we need to estimate only the variance of any one observation to know the other variances. We know that the squared deviation of each observation from the overall mean estimates its variance, so that if the first observation is twice as accurate as the rest its squared deviation must be multiplied by two in the estimated variance *i.e.* we take

$$[2(x_1-\bar{x})^2+(x_2-\bar{x})^2+(x_3-\bar{x})^2+\ldots+(x_n-\bar{x})^2]/(n-1)$$

where

$$\bar{x}=\frac{2x_1+x_2+\ldots+x_n}{2+1+\ldots+1}$$

as an estimate of the variance s^2, of the observations x_2, $x_3, \ldots x_n$ when $s^2/2$ is the estimated variance of x_1. Similarly, if the relative accuracies of the terms x_1, x_2, x_3 are, say, $1:2:3:\ldots$ then the estimated variance s^2 is

$$[(x_1-\bar{x})^2+2(x_2-\bar{x})^2+3(x_3-\bar{x})^2+\ldots]/(n-1)$$

where

$$\bar{x}=\frac{x_1+2x_2+3x_3+\ldots}{1+2+3+\ldots}$$

and the estimated variances of x_1, x_2, $x_3 \ldots$ are s^2, $s^2/2$, $s^2/3 \ldots$ To see how this works in practice consider the following example:

In an experiment to compare the egg-laying powers of birds on two diets, the birds were kept in runs in which the individual performances could not be judged. To overcome this difficulty the numbers of birds in each run were equalized, but the deaths of some of the birds gave, for a two week period, the results shown in the following table:

Run	Diet A					Diet B				
	1	*5*	*6*	*8*	*Total*	*2*	*3*	*4*	*7*	*Total*
No of birds	6	4	5	6	21	3	6	4	6	19
Eggs laid	55	41	49	63	208	29	49	29	54	161
Eggs/bird	9·17	10·25	9·80	10·50	9·90	9·67	8·17	7·25	9·00	8·47

The overall means have to be compared here, and the estimated variance of the egg production per bird on diet *A* is

$$[\,6\,(9{\cdot}17-9{\cdot}90)^2+4\,(10{\cdot}25-9{\cdot}90)^2+5\,(9{\cdot}80-9{\cdot}90)^2+6\,(10{\cdot}50-9{\cdot}90)^2\,]\;13$$

with 3 degrees of freedom. This can also be written

$$[\,6\times 9{\cdot}17^2+4\times 10{\cdot}25^2+5\times 9{\cdot}80^2+6\times 10{\cdot}50^2-208^2/21\,]13=6{\cdot}2929/3=2{\cdot}0976$$

by the rule given in section 1.10.

Similarly the estimated variance of the egg production per bird on diet *B* is

$$[\,3\times 9{\cdot}67^2+6\times 8{\cdot}17^2+4\times 7{\cdot}25^2+6\times 9{\cdot}00^2-161^2/19\,]13=13{\cdot}0069/3=4{\cdot}3356$$

with 3 degrees of freedom. This may be pooled with the other estimate to give an estimated variance $(6{\cdot}2929+13{\cdot}0069)/6=19{\cdot}2998/6=3{\cdot}2166$ with 6 degrees of freedom.

The standard error of the difference between the means is now $\sqrt{[\,3{\cdot}2166\,(1/21+1/19)\,]}=0{\cdot}568$, and the estimated deviate t is $(9{\cdot}90-8{\cdot}47)/0{\cdot}568=2{\cdot}52$. Referring to *Table V*, with 6 degrees of freedom we see that the deviate 2·45 is exceeded in 5 per cent of trials, so that this value of 2·52 will not occur very frequently by pure chance, and we conclude that this difference is probably real.

3A.13 *Proof of formula for variance of sum of independent measurements*—In section 3.5, it was stated that 'the variance of the sum or difference of a series of independent measurements is the sum of their variances'. This fundamental theorem can be proved as follows.

Obviously if it is true for two measurements it is true for any number, since we can add a further measurement to the sum of the two measurements and so on. Suppose we have two measurements, with deviations d_1 and d_2 from their mean and with variances $\sigma_1{}^2$ and $\sigma_2{}^2$. Then, by definition, $\sigma_1{}^2$ is the mean value of $d_1{}^2$ and $\sigma_2{}^2$ is the mean value of $d_2{}^2$. The deviation of the sum of the measurements will be d_1+d_2, and likewise its variance is the mean value of $(d_1+d_2)^2=d_1{}^2+2\,d_1\,d_2+d_2{}^2$. Now, since the deviations d_1 and d_2 may be positive or negative, d_1d_2 is equally likely to be positive or negative and its mean value is zero. Thus the mean value of $d_1{}^2+2\,d_1\,d_2+d_2{}^2$ is $\sigma_1{}^2+0+\sigma_2{}^2$, and the variance of the sum of the measurements is $\sigma_1{}^2+\sigma_2{}^2$. Similarly the variance of the difference between the two measurements is the mean value of $(d_1-d_2)^2=d_1{}^2-2\,d_1\,d_2+d_2{}^2$, which is again $\sigma_1{}^2+\sigma_2{}^2$.

It is now possible to prove that the use of the divisor $n-1$ in estimating the variance by the formula $\Sigma\,(x-\bar{x})^2/(n-1)$ is justified, but it must be remembered that the variance is the average value of the squared deviation of an observation from its true mean. Thus, if the true mean μ were known, $\Sigma(x-\mu)^2$ could be used to estimate $n\sigma^2$. Also, since Σx has a mean value of $n\mu$ and a variance of $n\sigma^2$, the average or expected value of $(\Sigma x - n\mu)^2$ is $n\sigma^2$. Now, using the formula of section 1A.14 with $a=\mu$

$$\Sigma(x-\bar{x})^2 = \Sigma(x-\mu)^2 - [\Sigma(x-\mu)]^2/n$$
$$= \Sigma(x-\mu)^2 - [\Sigma x - n\mu]^2/n$$

The average or expected value of the right hand side of the equation is $n\sigma^2 - n\sigma^2/n = (n-1)\sigma^2$, so that the average value of $\Sigma(x-\bar{x})^2/(n-1)$ is σ^2, as required.

SUMMARY OF PP 48 TO 53

It has been shown how it is possible to gauge the number of observations necessary to achieve any required degree of accuracy in the estimation of arithmetic means, or in their comparison.

The use of the inverse variance as an index of the relative reliability of measurement and its use in the combination of experimental results has been demonstrated.

The method of analysis to be applied when measurements are of differing precision has been discussed, and necessary alterations in the normal form of analysis have been pointed out.

EXAMPLES

27 In example *23*. show that 136 cores would have to be taken if it was desired to estimate the wireworm population to within 20 per cent with the same degree of accuracy.

28 25 male and 25 female students were given a psychological test of arithmetical ability. The mean scores and pooled variance of the two groups were 104·3, 98·2 and 159·3. Show that t is $6{\cdot}10/3{\cdot}57 = 1{\cdot}71$ and does not indicate a significant difference. If the difference between the two groups is real, find the least number of students that must be taken to be 99 per cent certain of its reality. *Ans*: $n = 318{\cdot}6\ (2{\cdot}58/6{\cdot}10)^2 = 57$.

29 Three observers each estimate the distance of Sirius by a series of observations. Their estimated distances are 7·94, 9·26 and 8·62 light years with standard errors of 0·61, 0·33 and 0·28 light years respectively. Show that the best combined estimate is 8·78 light years with a standard error of 0·20.

30 The numbers of ticks attaching themselves to sheep are removed and counted at irregular intervals. In a 3-day period the counts are 7, 12, 10, 4, 13, 9 and 8 and in the following 4-day period, the counts are 4, 14, 9, 10, 6, 8 and 12. Show that the mean daily attachment is 3·00 in the first period and 2·25 in the second period, and that the pooled standard deviation of this daily count is 1·74, with 12 degrees of freedom. Hence show that the standard error of the difference between means is $1{\cdot}74\sqrt{(1/21 + 1/28)} = 0{\cdot}50$.

4

COMPARISON OF SEVERAL SETS OF MEASUREMENTS

4.1 *The problem and its solution*—At first sight it appears that having solved the problem of comparing two sets of measurements we have also solved the problem of comparing several sets of measurements, but while it is possible to compare the sets in pairs, this course of action presents numerous difficulties. The first difficulty is that a large number of comparisons between pairs may be involved and consequently calculation may become excessive. For example, five sets of observations involve 10 comparisons between pairs and ten sets involve 45 comparisons. Secondly, if we make a large number of comparisons we shall expect to get at least one large difference by pure chance. The third difficulty occurs mainly in experimentation, but is fairly common in surveys and other work. In the comparison of means we ensure that a difference cannot be due to chance by estimating the variability in the population sample, and hence the variability of the means and their differences. If several means are to be compared then many observations must be taken and, as explained in section 3.7, these observations have to be randomly ordered. Consequently a greater period of time or a larger area will be covered, or more material will be used, and usually this will increase the variability and mask any true differences. For example, to compare two varieties of wheat each in, say, four $\frac{1}{40}$-acre plots covers only $\frac{1}{5}$ acre, and the area will probably not be very variable, but if ten varieties are compared, an acre of land must be used. This greater area will lead to a greater variability in the fertility and tend to obscure the differences. Fortunately it will be found that the solution of the first two difficulties leads to a method of overcoming this third difficulty.

The main problem is to find a method of testing several sets of observations simultaneously, and such a method can be derived using the variance-ratio test. Consider the four sets of observations 7, 8, 8, 10, 12; 5, 5, 6, 6, 8; 6, 7, 8, 9, 10; 5, 7, 7, 8, 8. The means of these sets are 9, 6, 8, and 7, and the estimated variances of each set are 16/4, 6/4, 10/4, 6/4 with 4 degrees of freedom. The pooled variance is thus

$$\frac{16+6+10+6}{4+4+4+4}=\frac{38}{16}=2{\cdot}375$$

with 16 degrees of freedom. This estimate indicates the variability within each group and does not assume that the true group means are equal.

If, however, we assume that the treatments have had no effect or that there is no difference between the true group means, we can use the group means to indicate the variability of the individual observations*. Thus the overall mean is 7·5 and the estimated variance of the group means 9, 6, 8, and 7 is

$$[(9-7{\cdot}5)^2+(6-7{\cdot}5)^2+(8-7{\cdot}5)^2+(7-7{\cdot}5)^2]/3=5/3=1{\cdot}667$$

and since the variance of the individual observations should be five (the number of observations in each group) times the variance of the means, we get a second estimate of the variance (from between the groups) as 25/3 = 8·333 with 3 degrees of freedom. These two values 2·375 and 8·333 both estimate the variance of the individual observations; the former from the variability within the groups and the latter from the variability between the groups. Any significant discrepancy between them can be due only to the assumption made in calculating the latter variance, namely that there is no real difference between the group means. If we test the ratio of these variances, 8·333/2·375 = 3·51, by entering *Table III* with 3 and 16 degrees of freedom, we find that as high a value as 3·24 occurs only in 5 per cent of trials by pure chance. Thus we conclude that there is a difference between the variances which can be ascribed to real differences between the groups.

It must be noted that the basis of this test is a comparison of the variability within groups with that between groups. In fact, if we consider the overall variance of the twenty observations, we get

$$\frac{1}{19}[(7-7{\cdot}5)^2+(8-7{\cdot}5)^2+(8-7{\cdot}5)^2+\ \dots\ +(8-7{\cdot}5)^2]=\frac{63}{19}$$

and this may be derived by pooling the two variances, 38/16 and 25/3, as described in section 3.4. Thus we have effectively partitioned the total variation of the whole set of observations into two portions, the variation within the groups and the variation between the groups, and we test the differences between groups by a comparison of these two portions.

4.2 *Analysis of variance*—This procedure is set out formally as follows:

Formal Analysis of Variance

	Degrees of freedom	*Sum of squares of deviations*	*Estimated variance, or mean square*	*Variance ratio*
Variation between groups	3	25	8·333	3·51
Variation within groups	16	38	2·375	
Total variation	19	63		

* As in section 3A.12 or in the same manner as the variance of the means in *Figure 18* might be used to estimate the variance of the individual observations.

The term for total variation is always included since it is quicker to subtract the variation between groups from the total variation than to calculate the variation within groups.

The wording in an analysis of variance is usually abbreviated to read:

Abbreviated Analysis of Variance

	D.f.	*S.s.*	*M.s.*	*V.r.*
Between groups	3	25	8·333	3·51
Within groups	16	38	2·375	
Total	19	63		

This abbreviated notation will be used in subsequent examples.

The technique of carrying out this test has been formulated so that it can be carried out with the minimum of computation. If $x_1, x_2, \ldots x_{n1}$; $y_1, y_2, \ldots y_{n2}$; ... are m groups of observations and the numbers of observations in the groups are $n_1, n_2, \ldots$ then we have to calculate the sums of squares for each group, Σx^2, $\Sigma y^2, \ldots$ the overall sum of squares $\Sigma x^2 + \Sigma y^2 + \ldots$ the correction terms for each group $(\Sigma x)^2/n_1$, $(\Sigma y)^2/n_2, \ldots$ and the overall correction term $\dfrac{(\Sigma x + \Sigma y + \ldots)^2}{n_1 + n_2 + \ldots} = \dfrac{T^2}{N}$, if N is the total number of observations and T is their total. These are then put in the analysis of variance as follows:

	D.f.	*S.s.*
Between groups	$m-1$	Sum of correction terms − overall correction term *i.e.* $\dfrac{(\Sigma x)^2}{n_1} + \dfrac{(\Sigma y)^2}{n_2} + \ldots - \dfrac{T^2}{N}$
Within groups	$N-m$	Overall sum of squares − sum of correction terms *i.e.* $\Sigma x^2 + \Sigma y^2 + \ldots - \dfrac{(\Sigma x)^2}{n_1} - \dfrac{(\Sigma y)^2}{n_2} - \ldots$
Total	$N-1$	Overall sum of squares − overall correction term *i.e.* $\Sigma x^2 + \Sigma y^2 + \ldots - \dfrac{T^2}{N}$

The estimated variances are, of course, found by dividing the sums of squares by the corresponding degrees of freedom. It should also be remembered that the observations can be reduced by a constant quantity to shorten the calculation.

Suppose we consider this method on the above example, first using the unreduced observations:

Group	*Observations*	*No. in group*	*Sum*	*Sum of squares*	*Correction term* = $\frac{(Sum)^2}{No.\ in\ group}$	*Group mean*
1	7, 8, 8, 10, 12	5	45	421	405	9
2	5, 5, 6, 6, 8	5	30	186	180	6
3	6, 7, 8, 9, 10	5	40	330	320	8
4	5, 7, 7, 8, 8	5	35	251	245	7
Total		20	150	1,188	1,150	

Overall correction term $=150^2/20=1{,}125$

The analysis of variance is, as before:

	D.f.	*S.s.*
Between groups	3	$1{,}150-1{,}125=25$
Within groups	16	$1{,}188-1{,}150=38$
Total	19	$1{,}188-1{,}125=63$

Alternatively, if each observation is reduced by, say, 7 the calculation is reduced to

Group	*Reduced observations*	*No. in group*	*Sum*	*Sum of squares*	*Correction term* = $\frac{(Sum)^2}{No.\ in\ group}$	*Reduced group mean*
1	0, 1, 1, 3, 5	5	10	36	20	2
2	−2, −2, −1, −1, 1	5	−5	11	5	−1
3	−1, 0, 1, 2, 3	5	5	15	5	1
4	−2, 0, 0, 1, 1	5	0	6	0	0
Total		20	10	68	30	

Overall correction term $=10^2/20=5$

	D.f.	*S.s.*
Between groups	3	$30-5=25$
Within groups	16	$68-30=38$
Total	19	$68-5=63$

By either method the group means can be obtained by dividing the group sums by the numbers of observations (remembering to add 7 for the latter method), and these can be compared in pairs if necessary. The standard error of the difference between means will be $\sqrt{[2{\cdot}375(1/5+1/5)]}=0{\cdot}97$, with 16 degrees of freedom, so that here we would conclude that the group differences as a whole are significant, but that the differences between groups 1 and 3, 2 and 4, and 3 and 4 are not significant.

It should be noted that this method provides another test for the difference between two sets of observations. The two tests are in fact equivalent since, when only two sets of observations are involved, the variance ratio is equal to the square of the estimated deviate t.

4·3 *Further examples of analysis of variance*—Since the analysis of variance is an important statistical technique the analytical procedure should be understood and appreciated. The following examples of the analysis of variance will show its wide scope.

Variety	*Yields oz*			
1	182,	214,	216,	231
2	196,	202,	208,	224
3	203,	212,	221,	242
4	198,	203,	207,	222
5	171,	192,	197,	204
6	194,	218,	223,	232
7	208,	216,	218,	239
8	183,	188,	193,	198

a Four 1/60-acre plots of each of eight varieties of oats were laid down at random, and the adjoining figures give the yields from each plot.

Obviously the calculation may be shortened if these are reduced by 200 and this has been done in the following analysis:

Variety	*Yield − 200*				*No. of plots*	*Sum*	*Sum of squares*	*Correction term*	*Group mean − 200*
1	−18,	14,	16,	31	4	43	1,737	462·25	10·75
2	−4,	2,	8,	24	4	30	660	225·00	7·50
3	3,	12,	21,	42	4	78	2,358	1,521·00	19·50
4	−2,	3,	7,	22	4	30	546	225·00	7·50
5	−29,	−8,	−3,	4	4	−36	930	324·00	9·00
6	−6,	18,	23,	32	4	67	1,913	1,122·25	16·75
7	8,	16,	18,	39	4	81	2,165	1,640·25	20·25
8	−17,	−12,	−7,	−2	4	38	486	361·00	−9·50
Total					32	255	10,795	5,880·75	

Overall correction term $=(255)^2/32=2{,}032{\cdot}03$.

The analysis of variance now becomes:

	D.f.	*S.s.*	*M.s.*	*V.r.*
Between varieties	7	5,880·75 − 2,032·03 = 3,848·72	3,848·72/7 = 549·82	549·82/204·76 = 2·69
Within varieties	24	10,795·00 − 5,880·75 = 4,914·25	4,914·25/24 = 204·76	
Total	31	10,795·00 − 2,032·03 = 8,762·97		

The variance ratio just exceeds the 5 per cent point, 2·42, and consequently the differences between the varieties are probably real. We can thus present the mean yields in a table:

Variety	7	3	6	1	2	4	5	8
Mean yield	220·25	219·50	216·75	210·75	207·50	207·50	191·00	190·50

Standard error of difference $=\sqrt{[204{\cdot}76\,(1/4+1/4)]}=\pm 10{\cdot}12$.

Referring to *Table V* with 24 degrees of freedom, it can be seen that a difference of $2{\cdot}06 \times 10{\cdot}12 = 20{\cdot}85$ would be exceeded by chance in only 5 per cent of trials. Thus we conclude that the only significant differences are those between the three best yielders, *7, 3* and *6,* and the two worst yielders, *5* and *8.*

b In a pre-war nutrition survey, 13 families (each containing a medium worker, his wife and two children under 6 years, and each spending comparable amounts on food) were classified according to whether they lived in a heavy industrial town, a light industrial town, or in a rural district. The following figures give the average daily protein consumption: heavy industrial, 209, 268, 244, 281, 253, 212; light industrial, 262, 275, 272, 231; rural, 331, 401, 351. These may be reduced by 250 to simplify the analysis, which now takes the form:

District	*Reduced observations*	*No. in group*	*Sum*	*Sum of squares*	*Correction term*	*Reduced group mean*
Heavy ind.	−41, 18, −6, 31, 3, −38	6	−33	4,455	182	−5·5
Light ind.	12, 25, 22, −19	4	40	1,614	400	10·0
Rural	81, 151, 101	3	333	39,563	36,963	111·0
	Total	13	340	45,632	37,545	

Overall correction term = 8,892.

	D.f.	*S.s.*	*M.s.*	*V.r.*
Between districts	2	37,545 − 8,892 = 28,653	14,326·5	17·7
Within districts	10	45,632 − 37,545 = 8,087	808·7	
Total	12	45,632 − 8,892 = 36,740		

The variance ratio, 17·7, greatly exceeds the value of 7·56 which occurs by chance in 1 per cent of trials (see *Table IV* with 2 and 10 degrees of freedom) so that the differences between districts are highly significant. The means are 244·5, 260·0 and 361·0, and the standard error of the difference between heavy and light industrial districts is $\sqrt{[808 \cdot 7\,(1/6+1/4)]} = \pm 18 \cdot 4$, so that this difference (which is less than its standard error) cannot be regarded as significant.

c The following table gives another example of the analysis of variance in which there are five groups of four observations.

Group	*Observations*	*No. in group*	*Sum*	*Sum of squares*	*Correction term*	*Group mean*
1	5, 5, 6, 7	4	23	135	132·25	5·75
2	5, 7, 7, 8	4	27	187	182·25	6·75
3	6, 7, 8, 8	4	29	213	210·25	7·25
4	6, 8, 9, 10	4	33	281	272·25	8·25
5	8, 8, 10, 12	4	38	372	361·00	9·50
	Total	20	150	1,188	1,158·00	

Overall correction term = 1,125.

	D.f.	*S.s.*	*M.s.*	*V.r.*
Between groups	4	1,158 − 1,125 = 33	8·25	4·125
Within groups	15	1,188 − 1,158 = 30	2·00	
Total	19	1,188 − 1,125 = 63		

The variance ratio is again larger than would normally occur by chance (only once in twenty times) and the individual differences between the means can be tested using the standard error $\sqrt{[2 \cdot 00\,(1/4+1/4)]} = \pm 1 \cdot 00$.

4.4 *Orthogonality and interaction of effects*—It frequently occurs that in comparing a set of factors it is impossible to hold all other factors constant, so that unless suitable precautions are taken the accuracy of the comparisons may be greatly reduced. For example, if we are comparing a series of medical, economic, or psychological measurements on people to determine the effect of age, these comparisons may be complicated by the effect of sex. Thus, if old men are compared with young women, any difference may be due to sex or age. The only manner in which we can assume that sex does not enter into the comparison is to choose the same proportion of each sex in each age group. The effect of sex is then said to be 'orthogonal' to the effect of age, since the former effect will not enter into the estimation of the latter, and *vice versa*.

It must be noted that the orthogonality of two effects implies that one can be estimated without the other entering into the estimation, but it does not imply that the two effects are independent of each other. Indeed, it is quite likely that the effect of age may be different for males and females, so that although it is possible to estimate the effect of age independently of sex (say, in a group, half male and half female) the difference between the effects of age for the two sexes is necessary for a full understanding of the effects. This difference is called the 'interaction' of the two effects.

Interactions are often as important as direct effects. The classical example of the importance of interactions is that of the fertilizer experiment. If we are considering the effect of nitrate and of potash, their individual applications may increase the yield of potatoes by one and three tons per acre, but the important problem is to know whether their joint application will increase the yield by more or less than four tons per acre*. The amount by which this increase differs from four tons per acre measures the interaction of the two fertilizers. It is also a measure of the difference between the effects of nitrate in the presence and absence of potash or, alternatively, the difference between the effects of potash in the presence and absence of nitrate. Interactions will be further considered in sections 4A.10 to 4A.13.

4.5 *Randomized block analysis*—It is obvious that if we wish to estimate an effect accurately other influencing effects should, if possible, be made orthogonal to the effect to be estimated. For example, if the difference between the sexes is to be estimated equal numbers of each sex should be drawn from each age group so that age cannot enter into the comparison, and if, say, weight is likely to be an influencing factor equal numbers of each sex should be drawn from each weight group. Likewise, since fertility

* Both of these are quite likely, since it may be considered that nitrate and potash act as 'food and drink to the plant, neither being much use on its own but acting doubly well in the presence of the other. Alternatively, it may be considered that the two 'tonics' may not act in unison, any more than two dinners eaten simultaneously would.

is an important factor in agricultural experiments the area should first be divided into a number of blocks, each as uniform as possible, and equal numbers of plots (usually one) of each treatment or variety randomly located in each block. This design, which is called the Randomized Block design, has the advantage that all comparisons are effectively made within blocks, since block differences are orthogonal to treatment effects. The randomized block design may also be applied to eliminate large variations in material or time. Thus, successive batches of observational material or successive periods of time might be regarded as blocks.

The idea that all factors, other than those under investigation, should be made as similar as possible in any comparison is one of the fundamentals of experimentation and can hardly be regarded as a novelty. However, the randomized block approach represents an improvement in that it is now possible to estimate and eliminate analytically these other factors if they are orthogonal to the main comparisons. For example, consider the observations employed in section 4.2 and suppose that the five observations in each group are taken on five successive days. Then 7, 5, 6, 5 are the observations taken on the first day, and so on.

The observations may be represented in the form:

Group\Day	*1*	*2*	*3*	*4*	*5*
1	7	8	8	10	12
2	5	5	6	6	8
3	6	7	8	9	10
4	5	7	7	8	8

Analysis of variances can be constructed to test the differences between groups or between days (for construction of the latter analysis, see example *c*, section 4.3) as follows:

	D.f.	*S.s.*
Between groups	3	25
Within groups	16	38
Total	19	63

	D.f.	*S.s.*
Between days	4	33
Within days	15	30
Total	19	63

Since the group-to-group and day-to-day effects are orthogonal (each group has one observation each day), it is possible to remove both group and daily variations from the total, giving an analysis of variance:

	D.f.	*S.s.*	*M.s.*	*V.r.*
Between groups	3	25	8·333	20·0
Between days	4	33	8·250	19·8
Residual variation	12	5	0·417	
Total	19	63		

It is seen that the day-to-day and group variations account for nearly all of the variation, and that by removing the two sets of comparisons simultaneously we determine the accuracy of both more exactly. The high variance ratio now leaves no doubt about the significance of the group or the day-to-day variation. It must however be emphasized that this technique can only be applied when the comparisons are orthogonal since otherwise the items in the analysis of variance are no longer independent of each other.

4.6 *Examples of randomized block analysis*—The following examples demonstrate further uses of the randomized block analysis.

a A series of experiments was carried out on the keeping quality of milk stored by three different methods while in transit. Owing to the initial variations in the quality of the milk, samples of milk from the same farm were stored by the three methods, and the difference between farms eliminated by a randomized block analysis. The series was carried out over several days so that because of the large daily changes in keeping quality caused by temperature it was also necessary to eliminate differences between days. Thus all differences between sets of samples were eliminated:

The keeping qualities of different samples in hours and the form of analysis are set out as below*.

Sample	*Keeping quality by method* *A*	*B*	*C*	*Sum*	*Sum of squares*	*Correction term* = $(Sum)^2/3$
1	18·5	17·0	18·0	53·5	955·25	954·08
2	15·0	15·5	16·0	46·5	721·25	720·75
3	31·5	32·0	33·0	96·5	3,105·25	3,104·08
4	29·5	28·5	29·5	87·5	2,552·75	2,552·08
5	17·0	16·0	16·5	49·5	817·25	816·75
6	12·0	12·0	12·0	36·0	432·00	432·00
7	29·0	29·5	29·5	88·0	2,581·50	2,581·33
8	28·0	26·5	27·5	82·0	2,242·50	2,241·33
9	20·5	20·0	21·0	61·5	1,261·25	1,260·75
10	16·0	15·5	17·0	48·5	785·25	784·08
11	33·0	33·5	32·5	99·0	3,267·50	3,267·00
12	31·0	30·0	31·0	92·0	2,822·00	2,821·33
Sum	281·0	276·0	283·5	840·5	21,543·75	21,535·56
Correction term = $Sum^2/12$	6,580·08	6,348·00	6,697·69			
	19,625·77 (total of A, B, C)					

Overall correction term = 19,623·34.

	D.f.	*S.s.*	*M.s.*	*V.r.*
Between samples	11	21,535·56 − 19,623·34 = 1,912·22	173·838	653·5
Between methods	2	19,625·77 − 19,623·34 = 2·43	1·215	4·64
Residual variation	22	1,920·41 − 1,912·22 − 2·43 = 5·76	0·262	
Total	35	21,543·75 − 19,623·34 = 1,920·41		

The variability in the keeping quality of samples taken on different farms at different times is very highly significant, and it is obvious that only by the elimination of this cause of variability could we have hoped to test the comparatively small differences between methods. The variance ratio testing the differences between methods is significant 'at the 5 per cent level' of 3·44, and the means for the three methods, 23·4, 23·0, and 23·6, show that the latter method has resulted in the best keeping

* In this analysis it is unnecessary to write down the sum of squares and correction terms for each sample if a calculating machine is available, and again it can be shortened by reducing each observation by, say, 20 hours.

quality. However, method C is not significantly different from method A since the standard error of the difference is $\sqrt{[0{\cdot}262\,(1/12+1/12)]} = \pm 0{\cdot}21$.

b In a sugar beet experiment, four blocks of six plots were used to compare the effects of six methods of cultivation. The yields in ton/acre are shown below:

Treatment \ *Block*	*1*	*2*	*3*	*4*	*Sum*
A	13·2	16·3	17·4	15·8	62·7
B	14·1	15·6	16·3	15·9	61·9
C	12·9	11·1	14·5	14·2	52·7
D	15·2	17·2	15·7	15·1	63·2
E	17·1	18·3	16·9	15·0	67·3
F	14·9	15·6	14·3	13·1	57·9
Sum	87·4	94·1	95·1	89·1	367·7

The analysis may again be simplified by reducing each observation by a value near the mean, say 15. We then get:

Treatment \ *Block*	*1*	*2*	*3*	*4*	*Sum*	*Sum of squares*	*Correction term*
A	−1·8	1·3	2·4	0·8	2·7	11·33	1·82
B	−0·9	0·6	1·3	0·9	1·9	3·67	0·90
C	−2·1	−3·9	−0·5	−0·8	−7·3	20·51	13·32
D	0·2	2·2	0·7	0·1	3·2	5·38	2·56
E	2·1	3·3	1·9	0·0	7·3	18·91	13·32
F	−0·1	0·6	−0·7	−1·9	−2·1	4·47	1·10
Sum	−2·6	4·1	5·1	−0·9	5·7	64·27	33·02
Correction term	1·13	2·80	4·33	0·14			
	8·40 (total)						

Overall correction term = 1·35.

	D.f.	*S.s.*	*M.s.*	*V.r.*
Between blocks	3	8·40 − 1·35 = 7·05	2·350	1·79
Between treatments	5	33·02 − 1·35 = 31·67	6·334	3·93
Residual	15	62·92 − 31·67 − 7·05 = 24·20	1·613	
Total	23	64·27 − 1·35 = 62·92		

The difference between treatments is significant 'at the 5 per cent level' 2·90, and although here the difference between blocks is not significant it is obvious that we have improved the accuracy of the experiment by removal of the variations due to blocks. The differences between pairs of treatment means can be tested using the standard error $\sqrt{[1{\cdot}613\,(1/4+1/4)]} = \pm 0{\cdot}90$.

4.7 *Latin and Graeco-Latin squares*—In a randomized block analysis the main causes of variation can be removed if they are orthogonal to the effects that are under investigation. This is normally achieved by taking equal numbers of observations in the groups to be compared under each set of conditions. Thus, in example *a* of the previous section each method was tried once under the prevailing daily and farm conditions, and in *b* each treatment was tested once in each block. However, it sometimes happens that this cannot be achieved since groups of factors may work to the exclusion of each other. As an example consider a man who is comparing the results obtained by, say, five experimental methods. He may be able to carry out five experiments during a day and to do this on several days, testing each method once on each day. He can thus eliminate day-to-day variations in his final analysis but his results will still be affected by variations throughout the day. This will not usually be important but changing conditions, such as an increase in temperature or a general improvement in technique, may influence the results so that it is desirable to distinguish between the early morning and late afternoon observations. It is usually impossible to carry out five experiments simultaneously and the obvious method of overcoming this difficulty is to arrange the five experiments so that over a 5-day period each experiment is carried out once at each of the five times, thus:

*Time**Day*	*1*	*2*	*3*	*4*	*5*
1	*A*	*B*	*C*	*D*	*E*
2	*D*	*C*	*E*	*B*	*A*
3	*C*	*A*	*D*	*E*	*B*
4	*B*	*E*	*A*	*C*	*D*
5	*E*	*D*	*B*	*A*	*C*

In this manner we ensure that differences due to the hour or day when the experiments are carried out are eliminated since both time and day are orthogonal to the experimental contrasts. This design, which is called a Latin square, is commonly used in agricultural experiments where each treatment is placed once in each row and once in each column so that row and column differences in fertility do not enter into the treatment comparisons.

It is possible to apply Latin squares to any number of treatments and the rows and columns of such squares might be used to eliminate differences of method or material, as well as spacial or temporal differences. The following table gives several examples of Latin squares*:

EXAMPLES OF LATIN SQUARES OF ORDERS 3–7

A	*B*	*C*
C	*A*	*B*
B	*C*	*A*

A	*B*	*C*	*D*
B	*A*	*D*	*C*
D	*C*	*B*	*A*
C	*D*	*A*	*B*

A	*B*	*C*	*D*	*E*
D	*A*	*E*	*C*	*B*
C	*D*	*B*	*E*	*A*
E	*C*	*A*	*B*	*D*
B	*E*	*D*	*A*	*C*

A	*B*	*C*	*D*	*E*	*F*
D	*E*	*A*	*B*	*F*	*C*
E	*C*	*D*	*F*	*B*	*A*
B	*A*	*F*	*E*	*C*	*D*
F	*D*	*E*	*C*	*A*	*B*
C	*F*	*B*	*A*	*D*	*E*

A	*B*	*C*	*D*	*E*	*F*	*G*
E	*D*	*B*	*C*	*A*	*G*	*F*
C	*F*	*G*	*B*	*D*	*A*	*E*
G	*A*	*F*	*E*	*C*	*B*	*D*
F	*C*	*D*	*A*	*G*	*E*	*B*
B	*E*	*A*	*G*	*F*	*D*	*C*
D	*G*	*E*	*F*	*B*	*C*	*A*

In order for the analysis to be strictly justified the treatments have to be assigned at random to the letters of a randomly chosen square.

It is possible to eliminate a third group of factors, if they are also chosen to be orthogonal to the treatments. For example, if the method is carried

* See FISHER, R. A. and YATES, F. *Statistical Tables for Biological, Agricultural and Medical Research* London, 1943, for examples of Latin squares for 2–12 treatments.

out on litters of five rats, then the experiment would be designed so that each method tests a rat from each litter. Thus, if a suffix denotes the litter from which the rat is drawn the following design might be used:

Time\Day	*1*	*2*	*3*	*4*	*5*
1	A_1	B_2	C_3	D_4	E_5
2	D_2	C_5	E_1	B_3	A_4
3	C_4	A_3	D_5	E_2	B_1
4	B_5	E_4	A_2	C_1	D_3
5	E_3	D_1	B_4	A_5	C_2

In this design one rat from each litter is tested on each day, at each time of day and by each method. A design such as this is called a Graeco-Latin square. Graeco-Latin squares do not exist for every number of treatments (none exists to test six treatments) and not all Latin squares may be turned into Graeco-Latin squares.

4.8 *Analysis of a Latin square*—It has been pointed out that the advantage of the Latin square design is that the row, column and main effects are all orthogonal to each other. Thus the estimation of any one set of effects will not be influenced by the other two sets and the items in the analysis of variance are all independent of each other. In carrying out the analysis the total variation is partitioned into the variations due to rows, columns, 'treatments', and the residual variation, which is used to test each of the other terms. The following example demonstrates the method of analysis.

In order to compare the bacterial counts of milk from five farms, samples were examined under a microscope on five days. Since several hours elapsed between the first and last samples counted the natural increase in the counts that occurred during this time was eliminated using the Latin square of section 4.7. For instance, farm *C* was tested at each time during the day. In the analysis the logarithms of the bacterial counts per cubic millimetre were used since changes tended to be proportional rather than absolute, *cf* section 2.3 and Chapter 8.

Analysis of the results proceeded as follows:

Time\Day	*1*	*2*	*3*	*4*	*5*	*Sum*	*Sum of squares*	*Correction term*
1	*A* 1·9	*B* 1·2	*C* 0·7	*D* 2·2	*E* 2·3	8·3	15·67	13·78
2	*D* 2·3	*C* 2·0	*E* 0·6	*B* 2·6	*A* 2·3	9·8	21·70	19·21
3	*C* 2·1	*A* 1·5	*D* 1·7	*E* 1·1	*B* 3·0	9·4	19·76	17·67
4	*B* 2·9	*E* 1·1	*A* 1·2	*C* 1·8	*D* 2·6	9·6	21·06	18·43
5	*E* 1·8	*D* 2·1	*B* 2·0	*A* 2·4	*C* 2·5	10·8	23·66	23·33
Sum	11·0	7·9	6·2	10·1	12·7	47·9	101·85	92·42
Correction term	24·20	12·48	7·69	20·40	32·36			
			97·03					
Farm	*A*	*B*	*C*	*D*	*E*			
Sum	9·3	11·7	9·1	10·9	6·9			
Correction term	17·30	27·38	16·56	23·76	9·52			
			94·52					

Overall correction term = 91·78.

	D.f.	*S.s.*	*M.s.*	*V.r.*
Between days	4	97·03 − 91·78 = 5·25	1·312	10·93
Between times	4	92·42 − 91·78 = 0·64	0·160	1·33
Between farms	4	94·52 − 91·78 = 2·74	0·685	5·71
Residual	12	10·07 − 5·25 − 0·64 − 2·74 = 1·44	0·120	
Total	24	101·85 − 91·78 = 10·07		

This analysis shows that removal of the day-to-day variation has caused a highly significant improvement in the accuracy of the comparison. Removal of the hourly variation has caused slight improvement in accuracy *i.e.* it may be a chance effect. The differences between farms is also seen to be significant and consequently we can test the differences between means using a standard error of $\sqrt{[0{\cdot}120\,(1/5+1/5)]} = \pm 0{\cdot}22$. The mean values are:

Farm	*A*	*B*	*C*	*D*	*E*
Mean	1·86	2·27	1·82	2·18	1·38

Since a difference of $0{\cdot}22 \times 2{\cdot}18 = 0{\cdot}48$ (where 2·18 is the value of t for 12 degrees of freedom) is exceeded by pure chance in 5 per cent of trials we conclude that the major difference is that between farm *E* and the other farms.

We should not combine the means from farms *A* and *C* or *B* and *D* unless we have an *a priori* reason for doing so. If however we can justify such a step, the means become:

Farm	*A + C*	*B + D*	*E*
Mean	1·840	2·26	1·380

The standard error of the difference between *E* and either of the other means is $\sqrt{[0{\cdot}120\,(1/5+1/10)]} = \pm 0{\cdot}19$, while the standard error of the difference between $A+C$ and $B+D$ is $\sqrt{[0{\cdot}120\,(1/10+1/10)]} = \pm 0{\cdot}15$, so that these three groups differ significantly from one another.

SUMMARY OF PP 54 TO 66

It has been shown that the variance-ratio test can be used as a method for comparing several sets of observations. This test can be presented formally in an analysis of variance table which partitions the total variability into two portions: the variability between the sets of observations and the variability within the sets of observations. Furthermore, if two sets of effects are observed in such a manner that neither enters into the estimation of the other they are said to be orthogonal. It is then possible to use an analysis of variance to partition the total variability into three portions: the variability due to either set of effects and the residual or 'unaccountable' variation. Both sets of effects may then be tested. This analysis is called the randomized block analysis.

Lastly, it has been shown how it is possible, using a Latin or Graeco-Latin square, to make three or four sets of effects orthogonal so that the analysis of variance can be partitioned into portions corresponding to each group of effects plus a residual portion.

EXAMPLES

31 Sections were taken from seven European larches of the same age at the same height from the ground, and four measurements of the trachoid length in mm were made from each aspect in each section. Using the values given below, show that the variation in trachoid length from tree to tree is greater than the variation within trees. Find the part of the total variability in trachoid length that can be ascribed to aspect and hence show that aspect does not have a significant effect on trachoid length.

Tree \ *Aspect*	*E*	*S*	*W*	*N*
1	3·4	3·5	3·1	3·5
2	2·8	3·1	3·0	3·0
3	3·0	3·2	3·3	3·3
4	3·0	3·0	2·5	2·8
5	3·3	3·5	3·7	3·6
6	3·3	3·0	2·9	2·8
7	3·4	3·6	3·7	3·6

32 Thirty students were given tests of intelligence and concentrative ability. The results were arranged in five groups according to concentrative ability as follows:

Concentrative ability	*Intelligence quotient*									
A	128,	120,	139							
B	131,	113,	114,	117,	122,	110				
C	115,	105,	131,	117,	129,	98,	120,	112,	110,	121
D	96,	103,	105,	104,	73,	102,	107,	95		
E	95,	87,	93							

Show that the differences in intelligence between the groups are not likely to be due to chance, using the analysis:

	D.f.	*S.s.*	*M.s.*	*V.r.*
Between groups	4	3,919	979·8	10·72
Within groups	25	2,286	91·4	
Total	29	6,205		

It should be noted that this analysis tests whether two variables are related. No assumption is made about the form of the relation since any two groups could be interchanged without affecting the final conclusion. For this reason this analysis is more general, but less sensitive, than one in which a particular form of relationship is assumed (such as in Chapter 6).

33 The statures of 18 pairs of brothers were measured to the nearest inch, and the following set of results was obtained:

65, 67; 69, 67; 70, 71; 67, 68; 71, 72; 68, 67; 66, 68; 70, 74; 65, 68; 68, 69; 70, 70; 70, 68; 72, 73; 65, 71; 66, 66; 69, 64; 69, 71; 67, 63.

Show that the following analysis of variance can be constructed:

	D.f.	*S.s.*	*M.s.*	*V.r.*
Between families	17	167	9·82	2·76
Within families	18	64	3·56	
Total	35	231		

and deduce that the statures of brothers tend to be similar.

34 In order to estimate and compare the growth of a seed mixture during three successive months, four plots of one square yard each were cut every month. The yields converted to lb/acre/day were:

Plot	*1*	*2*	*3*	*4*	*Mean*
April	16	13	15	9	13·25
May	30	22	29	23	26·00
June	24	20	18	24	21·50

Show that the analysis of variance of these figures is:

	D.f.	*S.s.*	*M.s.*	*V.r.*
Between plots	3	48	16·0	1·6
Between times	2	334	167·0	17·2
Residual	6	58	9·7	
Total	11	440		

and conclude that the standard error of the difference between means is $\pm 2{\cdot}20$.

35 Four blocks of five wheat plots infected with eye spot were used to test the control of the disease achieved by spraying with sulphuric acid at different stages of growth. One plot in each block was sprayed at each of four times, and the remaining plots remained unsprayed. The following figures give the estimated percentage of severely affected plants at harvest:

Time of spraying \ *Block*	*1*	*2*	*3*	*4*
None	42	25	46	33
November	31	21	27	37
February	14	18	23	12
March	6	20	14	27
April	44	31	29	36

Verify that the differences between times of spraying are significant and construct the following table to demonstrate the main conclusions.

Time of spraying	*None*	*November*	*February*	*March*	*April*
Mean percentage	36·50	29·00	16·75	16·75	35·00

Standard error of difference between two means $= \pm 5{\cdot}56$.
Significant difference at the 5 per cent level $= 12{\cdot}12$.

36 The quarterly averages of weekly freight car loadings in the U.S.A. for the period 1924-1928 are (in thousands):

Year \ *Quarter*	*1*	*2*	*3*	*4*	*Total*
1924	890	890	970	970	3,720
1925	920	970	1,050	1,010	3,950
1926	940	1,010	1,100	1,060	4,110
1927	970	1,000	1,050	970	3,990
1928	900	980	1,060	1,040	3,980
Total	4,620	4,850	5,230	5,050	19,750

Verify that there is a large seasonal variation using the analysis:

	D.f.	*S.s.*	*M.s.*	*V.r.*
Between years	4	20,250	5,062	7·2
Between quarters	3	41,335	13,778	19·7
Residual	12	8,390	699	
Total	19	69,975		

37 A 4×4 Latin square is used to compare four methods of cultivation on a potato crop. Using the following yields, which have been rounded off to the nearest ton/acre, verify that the methods have not given significantly different results:

D 12	*A* 11	*C* 13	*B* 12
B 14	*C* 13	*A* 12	*D* 15
A 10	*B* 9	*D* 10	*C* 11
C 12	*D* 13	*B* 11	*A* 12

Variance ratio for treatments = 2·125 with 3 and 6 degrees of freedom.

38 In an experiment to test the relative efficiencies of six foodstuffs, six litters, each containing three hogs and three gilts, were fed for a 12-week period. The three pigs of the same sex from any litter were arranged according to initial size and the six diets were given in Latin square formation, so that one pig from each litter received any particular diet. The weight gains over a period of three months were:

Pig \ Litter		1	2	3	4	5	6
Male	1	*A* 130	*B* 110	*C* 140	*D* 110	*E* 150	*F* 100
	2	*E* 130	*D* 90	*F* 130	*A* 100	*C* 140	*B* 90
	3	*C* 120	*A* 90	*B* 110	*E* 100	*F* 130	*D* 70
Female	1	*D* 100	*F* 120	*E* 140	*B* 110	*A* 120	*C* 110
	2	*F* 120	*E* 100	*D* 110	*C* 130	*B* 130	*A* 80
	3	*B* 110	*C* 100	*A* 120	*F* 120	*D* 140	*E* 100

Show that the analysis of these gains is:

	D.f.	*S.s.*	*M.s.*	*V.r.*
Sex and initial size	5	1,289	257·8	4·00
Litters	5	7,423	1,484·6	23·05
Diets	5	2,056	411·2	6·38
Residual	20	1,288	64·4	
Total	35	12,056		

and conclude that there is a significant difference between diets.

EXTENDED DEVELOPMENT

4A.9 *Testing particular comparisons*—Frequently, in comparing a set of treatments, certain comparisons are of more general interest than others. For example, if we test three levels of the same treatment, say a_1, a_2, a_3, then we are interested in the two increments $a_2 - a_1$ and $a_3 - a_2$ resulting from the increasing application of the treatment. More generally, we are interested in the sum of these increments $a_3 - a_1$ which gives the overall increment and their difference, $2a_2 - a_1 - a_3$, which gives the extent to which the increment decreases as the treatment level increases. This latter effect is normally called the curvature. Suppose we wish to make a test of the curvature when n observations are made at each treatment level and a_1, a_2, a_3 are the treatment means. The variance of $2a_2$ is then $4\sigma^2/n$ and the variances of a_1 and a_3 are both σ^2/n. Thus, by the theorem of section 3·5, the variance of $2a_2 - a_1 - a_3$ is $4\sigma^2/n + \sigma^2/n + \sigma^2/n = 6\sigma^2/n$. We may therefore test whether the curvature differs significantly from zero by using a t test. In general, any comparison of observations may be tested in this manner.

Any particular comparison may also be tested using the analysis of variance. To do this, it should be remembered that, if there is no effect, each degree of freedom in the analysis of variance should, on average, equal the mean square. Thus, in the above illustration, since the average value of $(2a_2 - a_1 - a_3)^2$ is $6\sigma^2/n$, the appropriate term in the analysis of variance is $n(2a_2 - a_1 - a_3)^2/6$. This has, on average, the value σ^2 if the true curvature is zero. The appropriate term in the analysis of variance may, in general, be calculated in this manner, subtracted from the treatment sum of squares, and tested in the ordinary manner.

Any comparison may be made in the manner described above, but if two or more comparisons are to be tested simultaneously in the analysis of variance, they must be orthogonal to one another. Tests of orthogonality and problems raised by non-orthogonality will be discussed later.

The following example will demonstrate how a particular effect may be tested:

In an experiment to compare the bone cavity areas of female rats after gestation and lactation, six groups of twenty rats were used. One group was killed and examined after each of the first, second and third lactations and gestations. Then the group killed after the first gestation may be denoted by G_1; after the first lactation by L_1, and so on. The mean cavity areas obtained from this experiment were:

G_1	L_1	G_2	L_2	G_3	L_3
2·72	3·47	3·26	3·96	3·57	3·97

Standard error of means = ±0·127

There is a marked tendency for the means to increase as the animals grow older, so that the direct comparison of $G_1+G_2+G_3$ with $L_1+L_2+L_3$ compares animals of different ages; the latter group being slightly older than the former. Alternatively, if we compare G_2+G_3 with L_1+L_2, this position is reversed. In consequence, the comparison of $\frac{1}{2}G_1+G_2+G_3$ with $L_1+L_2+\frac{1}{2}L_3$, will be relatively free from age effects*. Thus, we are interested in

$$2(\tfrac{1}{2}G_1+G_2+G_3-L_1-L_2-\tfrac{1}{2}L_3)/5=(G_1+2G_2+2G_3-2L_1-2L_2-L_3)/5=-0{\cdot}49$$

which measures the average difference between the gestation and lactation measurements. Here, the variance of G is the variance of the mean, and the variance of $2G$ is 4 times the variance of the mean. Therefore the variance of the estimated effect is $(1+4+4+4+4+1)/25=18/25$ths of the variance of the means and the standard error of this effect is $\pm 0{\cdot}127\sqrt{(18/25)}=\pm 0{\cdot}108$. It is, of course, highly significant.

4A.10 *Test for interaction*—It has been pointed out in section 4.4 that the interaction of two effects may be as important as the effects themselves. For instance, the interaction of nitrate and potash in agricultural practice *i.e.* the extent to which the application of nitrate will influence the effect of potash or *vice versa*, must largely determine the applications of each fertilizer that are economically justified. Likewise, the interaction of age and sex is a common occurrence in medical and psychological investigations.

Consider a simple experiment in which four field plots are given nitrate and potash nk, nitrate alone n, potash alone k, and no application (1), respectively. The results of this experiment may be tabulated in the form:

	No potash	*Potash*	*Total*
No nitrate	9	12	21
Nitrate	10	15	25
Total	19	27	46

* This is not the only possible comparison that may be made but it is one of the most efficient.

in which 15 ton/acre is the yield of the nitrate and potash plot, 12 ton/acre is the yield of the potash plot *etc.*

The total sum of squares for this experiment is $9^2+10^2+12^2+15^2-\frac{46^2}{4}=21$, while the sums of squares due to the effects of nitrate and potash are $\frac{21^2}{2}+\frac{25^2}{2}-\frac{46^2}{4}=4$, and $\frac{19^2}{2}+\frac{27^2}{2}-\frac{46^2}{4}=16$, respectively. The analysis of variance is thus:

	D.f.	*S.s.*	*M.s.*	*V.r.*
Nitrate effect	1	4	4	——
Potash effect	1	16	16	——
Residual	1	1	1	
Total	3	21		

Now the nitrate effect with one degree of freedom (it involves one comparison *i.e.* comparison of the nitrate with the no-nitrate plots) removes from the total variability the portion that might be ascribed to the nitrate, and likewise with the potash effect, so that the residual measures either the true variability or the extent to which the nitrate effect depends upon the level of potash *i.e.* the nitrate-potash interaction. Previously we have assumed that the interactions of effects were negligible and that the residual term measured the true variability. This is still true provided that the effects that are considered act independently of one another, but otherwise the residual term may be partly ascribed to interaction.

A useful alternative method of regarding this analysis is to consider three possible comparisons contributing to the total variability. The first comparison

$$nk+n-k-(1)=15+10-12-9=4$$

which is the difference between the nitrate and no-nitrate plots, indicates the effect of nitrate. The second comparison

$$nk-n+k-(1)=15-10+12-9=8$$

similarly indicates the effect of potash. The last comparison

$$nk-n-k+(1)=15-10-12+9=2$$

can also be written

$$[nk-k]-[n-(1)]$$

i.e. the difference between the effect of nitrate in the presence and the effect of nitrate in the absence of potash, or

$$[nk-n]-[k-(1)]$$

i.e. the difference between the effects of potash in the presence and absence of nitrate, and indicates the interaction of nitrate with potash. Furthermore, if each of these values is squared and divided by the number of plots *i.e.* four, the corresponding term in the analysis of variance is obtained. (This follows from the methods given in the previous section.) Thus $4^2/4=4$, $8^2/4=16$ and $2^2/4=1$, are the three terms in the analysis of variance. A consequence of the effect of interaction on the residual variation is that if the interaction of two effects is to be removed and tested each treatment combination must be repeated; replication is necessary if an estimate of the residual variation is to be obtained. Thus, in the above example, if we recognize that the residual is in fact an interaction of nitrate with potash the treatment combinations will have to be repeated if the interaction is to be tested.

Suppose a second set of observations is taken:

	No potash	*Potash*	*Total*
No nitrate	10	13	23
Nitrate	12	17	29
Total	22	30	52

When these are combined with the previous set, we get:

	No potash	*Potash*	*Total*
No nitrate	19	25	44
Nitrate	22	32	54
Total	41	57	98

The total sum of squares is now

$$9^2+10^2+12^2+15^2+10^2+12^2+13^2+17^2-\frac{98^2}{8}=51{\cdot}5$$

The sum of squares due to treatments is

$$\frac{19^2}{2}+\frac{22^2}{2}+\frac{25^2}{2}+\frac{32^2}{2}-\frac{98^2}{8}=46{\cdot}5$$

and included in this value are the sums of squares due to the effect of nitrate

$$\frac{44^2}{4}+\frac{54^2}{4}-\frac{98^2}{8}=12{\cdot}5$$

due to the effect of potash

$$\frac{41^2}{4}+\frac{57^2}{4}-\frac{98^2}{8}=32{\cdot}0$$

and due to the nitrate-potash interaction

$$46{\cdot}5 - 12{\cdot}5 - 32{\cdot}0 = 2{\cdot}0$$

The complete analysis of variance is now:

	D.f.	*S.s.*	*M.s.*	*V.r.*
Nitrate effect	1	12·5	12·5	10·0
Potash effect	1	32·0	32·0	25·6
Interaction	1	2·0	2·0	1·6
Treatments	3	46·5	——	——
Residual	4	5·0	1·25	
Total	7	51·5		

Using *Tables III* and *IV*, we conclude that the nitrate and potash differences are not likely to be due to chance (they are 5 and 1 per cent significant respectively) but we cannot conclude that the interaction of potash and nitrate is real *i.e.* that the application of nitrate affects the response to potash.

It is possible in the same manner as previously to remove the variability due to other causes provided that they are orthogonal to the treatments. Thus we can remove the variability due to the differences between the two groups or blocks of treatments, which is

$$\frac{46^2}{4} + \frac{52^2}{4} - \frac{98^2}{8} = 4{\cdot}5$$

and the revised analysis becomes:

	D.f.	*S.s.*	*M.s.*	*V.r.*
Nitrate effect	1	12·5	12·5	75·0
Potash effect	1	32·0	32·0	192·0
Interaction	1	2·0	2·0	12·0
Treatments	3	46·5	——	——
Between blocks	1	4·5	4·5	27·0
Residual	3	0·5	0·167	
Total	7	51·5		

It is seen that the majority of the original residual can be ascribed to block differences, and that we are now able to conclude that the interaction of nitrate and potash is significant at the 5 per cent level.

4A.11 *Examples of test for interaction*—The following examples show the applications of the test for interaction.

a In order to determine the effect of initiative and parental encouragement upon the intelligence of 11-year old children, 48 children were tested to determine their degree of initiative and extent of parental encouragement, and subsequent tests of intelligence gave the following intelligence quotients:

Init.	*Parental encrgmt*	*Intelligence quotients*	*Total*
High	*High*	107, 126, 122, 129, 117, 128, 103, 117, 132, 139, 122, 121	1,463
High	*Low*	99, 95, 79, 94, 122, 117, 99, 102, 110, 116, 121, 96	1,250
Low	*High*	86, 89, 96, 101, 81, 99, 113, 79, 89, 82, 91, 74	1,080
Low	*Low*	104, 107, 93, 92, 82, 87, 100, 80, 102, 103, 85, 69	1,104

The normal method of analysis (as described in section 4.2) gives the analysis of variance:

	D.f.	*S.s.*	*M.s.*
Between groups	3	7,744	2,581
Within groups	44	5,823	132
Total	47	13,567	

The totals of the high and low initiative groups are 2,713 and 2,184, so that the sum of squares due to initiative is

$$\frac{2{,}713^2}{24}+\frac{2{,}184^2}{24}-\frac{4{,}897^2}{48}=5{,}830$$

The corresponding sum of squares due to parental encouragement is

$$\frac{2{,}543^2}{24}+\frac{2{,}354^2}{24}-\frac{4{,}897^2}{48}=744$$

The completed analysis now becomes:

	D.f.	*S.s.*	*M.s.*	*V.r.*
Initiative	1	5,830	5,830	44·17
Parental encouragement	1	744	744	5·64
Interaction	1	1,170	1,170	8·86
Between groups	3	7,744	—	—
Within groups	44	5,823	132	
Total	47	13,567		

The analysis shows quite clearly that initiative is related to intelligence. Parental encouragement exceeds the '5 per cent significance level' 4·06, but here the importance lies in the high value of the interaction. This value verifies what is indicated by the means: parental encouragement is not effective for children with low initiative. The analysis may be completed by a table of the means:

	High initiative		*Low initiative*	
	High P.E.	*Low P.E.*	*High P.E.*	*Low P.E.*
Mean I.Q.	121·9	104·2	90·0	92·0

Standard error of the difference between two means is $\sqrt{[132\,(1/12+1/12)]}=4{\cdot}69$.

b In an agricultural experiment three varieties of barley were tested and simultaneously four levels of fertilizer were applied, so that twelve combinations of treatments were used. These twelve combinations were replicated in two blocks and the following yields of grain in cwt/acre were obtained:

Level \ *Variety*	*Block 1*				*Block 2*			
	1	*2*	*3*	*Total*	*1*	*2*	*3*	*Total*
1	14·4	11·8	13·0	39·2	13·9	12·1	12·6	38·6
2	12·6	13·3	14·8	40·7	15·8	13·2	15·6	44·6
3	16·5	13·4	14·4	44·3	15·3	14·8	15·1	45·2
4	15·1	14·3	13·6	43·0	15·5	14·6	14·5	44·6
Total	58·6	52·8	55·8	167·2	60·5	54·7	57·8	173·0

Level \ *Variety*	*Both blocks*			
	1	*2*	*3*	*Total*
1	28·3	23·9	25·6	77·8
2	28·4	26·5	30·4	85·3
3	31·8	28·2	29·5	89·5
4	30·6	28·9	28·1	87·6
Total	119·1	107·5	113·6	340·2

Here the interaction of levels with varieties is particularly important since a significant interaction would indicate that the optimum level was probably different for the three varieties. The uncorrected total sum of squares is 4,857·34 and the correction term is $(340\cdot2)^2/24=4{,}822\cdot34$, so that the corrected total sum of squares is 35·00. The sum of squares due to treatments is

$$\frac{(28\cdot3)^2}{2}+\frac{(28\cdot4)^2}{2}+\ldots+\frac{(28\cdot1)^2}{2}-\frac{(340\cdot2)^2}{24}=4{,}849\cdot17-4{,}822\cdot34=26\cdot83$$

and this may be subdivided into a portion due to varieties

$$\frac{(119\cdot1)^2}{8}+\frac{(107\cdot5)^2}{8}+\frac{(113\cdot6)^2}{8}-\frac{(340\cdot2)^2}{24}=4{,}830\cdot75-4{,}822\cdot34=8\cdot41$$

a portion due to levels of fertilizer

$$\frac{(77\cdot8)^2}{6}+\frac{(85\cdot3)^2}{6}+\frac{(89\cdot5)^2}{6}+\frac{(87\cdot6)^2}{6}-\frac{(340\cdot2)^2}{24}=4{,}835\cdot49-4{,}822\cdot34=13\cdot15$$

and a remaining portion due to the interaction of levels with varieties

$$26\cdot83-8\cdot41-13\cdot15=5\cdot27$$

Lastly, the sum of squares due to block differences is

$$\frac{(167\cdot2)^2}{12}+\frac{(173\cdot0)^2}{12}-\frac{(340\cdot2)^2}{24}=2\cdot80$$

The residual may be found by subtracting the blocks and treatments sums of squares from the total, and the complete analysis may now be presented:

	D.f.	*S.s.*	*M.s.*	*V.r.*
Between varieties	2	8·41	4·205	6·84
Between levels	3	13·15	4·383	7·12
Interaction	6	5·27	0·878	1.43
Between treatments	11	26·83	——	——
Between blocks	1	1·40		2·28
Residual	11	6·77	0·615	
Total	23	35·00		

There can be little doubt of the significance of the differences between varieties or levels, since their variance ratios are both near the 1 per cent values given in *Table IV*, but their interaction can equally obviously be ascribed to chance variability. Thus we conclude that there is no evidence to show that the optimum level of fertilizer is different for the three varieties. To complete the analysis, tables of means are required as follows:

Variety	*1*	*2*	*3*
Mean	14·89	13·44	14·20

Standard error of difference between two means $= \sqrt{[0{\cdot}615\,(1/8+1/8)]} = \pm 0{\cdot}39$.

Level	*1*	*2*	*3*	*4*
Mean	12·97	14·22	14·92	14·60

Standard error of difference between two means $= \sqrt{[0{\cdot}615\,(1/6+1/6)]} = \pm 0{\cdot}45$.

These tables show that although variety *1* has given the best yield it is not significantly better than variety *3*. Similarly, level *3* is not significantly better than levels *2* or *4*. Further experimentation would be required before the optimum variety and level could be ascertained, but at this stage it is possible to rule out variety *2* and level *1*.

c In order to compare the time taken to carry out two different tests, the tests were carried out in duplicate on six different occasions. The times (to the nearest hour) were:

*Test**Day*	*1*	*2*	*3*	*4*	*5*	*6*	*Total*
A	7	9	8	6	8	7	96
	9	10	8	8	9	7	
B	8	10	7	10	7	8	105
	9	11	8	10	8	9	
Total	33	40	31	34	32	31	201

and the completed analysis of variance was:

	D.f.	*S.s.*	*M.s.*	*V.r.*
Tests	1	3·375	3·375	5·40
Days	5	14·375	2·875	4·60
Tests × days	5	10·375	2·075	3·32
Groups	11	28·125	——	——
Residual	12	7·500	0·625	
Total	23	35·625		

The three variance ratios are all significant at the 5 per cent level. This shows first that the day-to-day variation is greater than the variation on any particular day, secondly that the relative performances of the two tests vary from day-to-day and thirdly that the overall performances of the two tests over the six days were significantly different. However, a further problem now arises: assuming that the six days are representative, can we conclude that the two tests will be significantly different over a period of time? In order to do so we must ensure that the difference between the tests could not have arisen as a result of the day-to-day variability *i.e.* that we have not been unfortunate in our days of testing. It must be noted that this is a different basis from the usual test in which we are concerned with determining whether tests are different under the existing conditions. Thus, if we wish to determine whether the tests will be significantly different over a period of time, the variability due to the tests must be compared with the day-to-day variability in their relative performance *i.e.* the tests × day interaction. This gives a variance ratio $3{\cdot}375/2{\cdot}075 = 1{\cdot}63$ which is not significant, so that we could not conclude that the two tests would differ on continued application.

It may now be observed that in the randomized block experiment the use of the treatment × block interaction as error is justified, even when the effect of treatments is likely to change from block to block since we normally wish to apply the results to a wider area than the limits of the experiment.

4A.12 *Higher order interactions and factorial principle*—In section 4A.9 it was shown that the variability in the results obtained from using all combinations of two sets of factors may be split into three parts: the variability due to either set of factors and the variability due to their interaction. In the same manner if we use all of the combinations of three sets of factors the total variability can be split into seven parts: the variability due to each set of factors (three parts), the variability due to the interactions of pairs of sets of factors (three parts) plus a residual part which is due to the joint interaction of the three sets of factors. This residual portion, which is called a second order interaction, might be regarded as indicating the change in the interaction of any two sets of factors when the third set is introduced.

To demonstrate this idea, consider an experiment in which the eight combinations of nitrate n, potash k and phosphate p are applied. These combinations are $n\,k\,p$, $n\,k$, $n\,p$, $k\,p$, n, k, p, and no fertilizer (1), and the seven degrees of freedom corresponding to the comparisons of these treatments can be apportioned as previously. Three of these correspond to main effects, such as the nitrate effect obtained by comparing the nitrate plots with the no-nitrate plots:

$$n\,k\,p+n\,k+n\,p+n-k\,p-k-p-(1)$$

A further three correspond to the direct or first order interactions, such as the nitrate-potash interaction obtained by comparing the nitrate effects in the presence and absence of potash:

$$[n\,k\,p+n\,k-k\,p-k]-[n\,p+n-p-(1)]$$

The remaining degree of freedom corresponds to the second order interaction, the nitrate-potash-phosphate interaction, obtained by comparing the nitrate-potash interactions in the presence and absence of phosphate:

$$[nkp-np-kp+p]-[nk-n-k+(1)]=nkp-np-kp-nk+n+k+p-(1)$$

As before, these expressions when squared and divided by eight determine the corresponding terms in the analysis of variance.

In the same manner, a third order interaction may be defined as the change in a second order interaction due to the introduction of a further set of factors but, in practice, it is seldom that interactions of the second and higher orders are significantly different from zero. These higher order interactions are important when the joint action of several sets of factors is required to produce an effect, but, in general, it may be assumed that they are negligible.

This assumption allows us to leave the terms corresponding to the higher order interactions in the residual of the analysis of variance. Thus, when several sets of factors are involved it is unnecessary to replicate the observations, since a measure of the natural variability is provided by these interactions. Thus, if all of the combinations of several sets of factors are used in an investigation, a measure of variability can be derived from the higher order interactions. Such an investigation is said to employ the factorial principle. Replication may, however, be necessary unless the higher order interactions provide sufficient degrees of freedom to derive an accurate estimate of the residual variability. Usually, at least five degrees of freedom are required to measure the residual variability if the values of the variance ratio and t needed for significance are not to be very large.

It is impossible in a short space to describe all of the devices that may be used in analysing a factorial design and reference should be made to books on experimental design*. The general principles are as described above, but the form of analysis may be carried out with a minimum of difficulty using rapid techniques, careful checks being employed at each stage of calculation. In this manner, the analysis of factorial design is so arranged that it can be carried out as a routine calculation without any great difficulty by the non-specialist.

4A.13 *Confounding*—The advantage of factorial design in experimentation is that several sets of factors and their interrelations can be investigated simultaneously. However, a factorial design will normally involve a large number of observations, which will use a great deal of time, space, or material, and consequently may increase the variability of the results. For example, an experiment with five factors each at two levels requires $2^5 = 32$ observations, while five factors at three levels involve $3^5 = 243$ observations. This is a serious defect, since an increase in variability tends to obscure any real effects and to nullify the advantage gained by the factorial design, but the difficulty may be overcome by arranging the observations in blocks or strata in such a manner that the differences between blocks are identified with comparisons of little importance. The accuracy of the remaining comparisons is consequently increased, since the variability due to blocks can now be removed.

* YATES, F. The Design and Analysis of Factorial Experiments *Commonwealth Agricultural Bureaux Publication, No. 35*, for example.

To demonstrate this idea, consider an experiment in which the eight combinations of three sets of factors at two levels are contained in two blocks: $n\,k\,p$, n, k, p, and $n\,k$, $n\,p$, $k\,p$, (1). The nitrate-potash-phosphate interaction, which is determined by the difference $n\,k\,p - n\,k - n\,p - k\,p + n + k + p - (1)$, is then equivalent to the comparison between blocks and is effectively lost since it cannot be distinguished from a difference between blocks. All other comparisons are orthogonal to blocks. For example, the nitrate-potash interaction is determined by the difference $n\,k\,p + n\,k - k\,p - k - n\,p - n + p + (1)$ (as in section 4A.12) which is the sum of the comparisons, $n\,k\,p + p - n - k$ and $n\,k + (1) - n\,p - k\,p$, from within each block. Normally at least eight observations are contained in a block, but the same principle can be extended to larger experiments.

This method of sacrificing certain comparisons in order to increase the accuracy of the remaining comparisons is called confounding, and the lost comparisons are said to be confounded.

It is impossible to describe briefly a general method of determining a design to maintain certain comparisons and confound others, and reference must be made elsewhere for details of such methods (see bibliography), although the above example demonstrates how an experiment involving sets of factors at two levels can be designed to confound one comparison. The analysis of such designs presents no new problems since it merely involves removing the sums of squares for blocks and the unconfounded treatment comparisons from the total sum of squares, as in previous analyses.

SUMMARY OF PP 69 TO 79

Methods have been given for testing particular comparisons. It has been shown that if two sets of effects are orthogonal the analysis of variance can be used to estimate and test the extent of their interaction. This concept has been extended so that the joint interaction of three or more sets of effects can also be estimated, and the use of the factorial design in investigations has been demonstrated.

The principle of confounding, by which certain comparisons are sacrificed to increase the accuracy of the remaining comparisons, has also been described.

EXAMPLES

39 In the example of section 4A.9 the average difference of rats at gestation and lactation may be estimated alternatively using the formula

$$(5G_1 - 11L_1 + 8G_2 - 8L_2 + 11G_3 - 5L_3)/24 = -0{\cdot}45$$

Show that the standard error of this estimate is

$$0{\cdot}127\sqrt{(35/48)} = \pm 0{\cdot}108$$

Hence this more complicated estimate is very little different from the previous estimate.

40 Three replicates of a factorial experiment testing the effects and interactions of nitrate *n*, phosphate *p*, and potash *k* upon the yield of hay were carried out in three blocks. The following results were obtained:

Treatment\Block	*I*	*II*	*III*
(1)	26·6	31·8	33·3
n	43·9	44·1	50·7
p	29·6	32·9	39·0
np	43·3	46·6	35·7
k	29·4	31·4	32·6
nk	43·2	35·5	44·1
pk	29·0	28·2	30·9
npk	42·7	45·1	42·5

Show that the analysis of variance appropriate to this experiment is:

	D.f.	*S.s.*	*Ms.*
Nitrate effect	1	848·47	
Phosphate effect	1	0·05	
Potash effect	1	21·85	
Nitrate-phosphate interaction	1	4·25	
Nitrate-potash interaction	1	0·01	
Phosphate-potash interaction	1	1·26	
Nitrate-phosphate-potash interaction	1	53·11	
Treatments	7	929·00	132·71
Blocks	2	28·41	14·20
Residual	14	192·18	13·73
Total	23	1,149·59	

Hence conclude that only the nitrate effect is significant.

41 The following figures give the total bone ash weights of rats raised on three different diets by two different methods:

Diet\Method	*I*	*II*
A	600·8	421·3
	588·6	466·2
	527·8	463·3
	519·3	368·8
	518·9	335·8
	468·6	384·9
B	544·2	369·4
	466·1	465·9
	440·6	378·2
	514·5	407·7
	535·4	404·5
	499·2	401·4
C	638·8	361·4
	592·6	397·0
	465·5	397·3
	469·1	394·0
	554·0	409·9
	471·5	340·9

Show that the analysis of variance of this experiment is:

	D.f.	*S.s.*	*M.s.*
Diets	2	2,505	1,252
Methods	1	140,325	140,325
Diets × *methods*	2	4,365	2,182
Treatments	5	147,195	29,439
Residual	30	70,925	2,364
Total	35	218,120	

and compile a table of means with standard errors to demonstrate the results of this experiment.

5

ATTRIBUTES AND COMPARISON OF PROPORTIONS

5.1 *Measurement of attributes*—So far we have mostly been considering quantitative measurements and, in particular, in Chapters 3 and 4, measurements which are normally distributed. On many occasions, however, observations are taken which are not expressed on a quantitative scale. Thus colour, taste, shape and roughness may be observed, but these are not easily expressed according to any scale of measurement. When this occurs, a large number of observations is required in each group before any comparisons can be made. Then it is possible to compare the proportions in each group with the given attributes. In particular, we may wish to compare 'all or nothing' measurements such as deaths and survivals.

A comparison of proportions based upon small numbers of observations is more difficult than a comparison based upon large numbers. Equally the comparison of small proportions is usually difficult unless sufficient observations are taken. In the next sections it will be assumed that each proportion is derived from at least ten observations and that no attribute is observed less than five times. Where this is not so it may be necessary either to group the attributes so that, for example, light brown and dark brown hair are classified under one heading or, alternatively, to carry out a more lengthy exact analysis of the type to be described later.

The advantage of assuming that each proportion is based upon a large number of observations is that it is then determined by a large number of independent results. Consequently, such proportions tend to be normally distributed as has already been observed in section 2A.6 and *Figures 14* and *15*. We may therefore apply the methods developed for the normal distribution to test observed proportions.

There is one outstanding difference between tests carried out on normally distributed measurements and tests carried out on proportions. If proportions are used it is possible to predict theoretically their variance, whereas it is usually necessary to estimate the variance from the observations with normally distributed measurements. Thus it can be shown that the estimates of a proportion, π, based upon n observations will be distributed about the mean, π, with variance $\pi(1-\pi)/n$. It is seen that $\pi(1-\pi)$ corresponds to the variance of a single observation in the ordinary analysis while $\pi(1-\pi)/n$ is the variance of the mean of n observations. This is particularly useful since it provides the residual mean square in the analysis of variance.

Since only an estimate of π is known it would seem that the variance is also unknown. However, if we replace the true proportion π by the observed proportion p the variance will usually be fairly accurate. For example, if $\pi=0{\cdot}5$ and 100 observations are taken with 40 successes, the observed proportion is 0·4. The true variance of each observation will be

$$0{\cdot}5\,(1-0{\cdot}5)=0{\cdot}25$$

but using the observed proportion it will be calculated as

$$0{\cdot}4\,(1-04)=0{\cdot}24$$

The difference is negligible.

The tests described in the following sections employ, therefore, two approximations:

1 the observed proportion is assumed to be normally distributed;

2 the replacement of the true proportion by the observed proportion in calculating the variance is assumed to have little effect.

Both of these approximations will be of little importance provided sufficient observations are taken, but some care should be taken to ensure that the derived tests are not invalidated for this reason.

5.2 *Comparison of several groups*—If we have several groups of observations and the proportions in each group with a certain attribute are observed, then we may wish to test whether these proportions differ significantly. This may be done in exactly the same manner as for ordinary measurements *i.e.* by the variance-ratio test. Let $n_1, n_2 \ldots n_r$ be the numbers of observations in the r groups and let N be the total number of observations. Suppose, further, that the numbers in each group with the given attribute are $m_1, m_2 \ldots m_r$ and that the overall number with the attribute is M, so that the overall proportion with the attribute is $p=M/N$. The sum of squares between groups is as usual

$$\frac{m_1{}^2}{n_1}+\frac{m_2{}^2}{n_2}+\ldots+\frac{m_r{}^2}{n_r}-\frac{M^2}{N}$$

with $r-1$ degrees of freedom. The mean square between groups is thus

$$\frac{1}{r-1}\left[\frac{m_1{}^2}{n_1}+\frac{m_2{}^2}{n_2}+\ldots+\frac{m_r{}^2}{n_r}-\frac{M^2}{N}\right]$$

and this has to be compared with $p\,(1-p)$ to test whether the variation in numbers with the attribute between the groups is greater than would be expected. For this purpose we use the variance ratio

$$\frac{1}{p\,(1-p)}\times\frac{1}{r-1}\left[\frac{m_1{}^2}{n_1}+\frac{m_2{}^2}{n_2}+\ldots+\frac{m_r{}^2}{n_r}-\frac{M^2}{N}\right]$$

with $r-1$ and an infinite number of degrees of freedom. The following example will serve to demonstrate this test.

The resistance to tubercular infection of three groups of mice fed on different diets was measured by their survivals after three weeks. The results of an experiment are shown below:

Diet	*Number of mice, n*	*Number of survivals, m*	*Proportion of survivals, m/n*	*m2/n*
A	43	10	0·233	2·32558
B	41	12	0·293	3·51220
C	40	3	0·075	0·22500
Total	124	25	0·20161	6·06278

Overall correction term $=(25)^2/124=5{\cdot}04032$.

It is desired to test the differences between the proportions surviving on each diet. The mean square between diets is measured by $(6{\cdot}06278-5{\cdot}04032)/2=0{\cdot}51123$. This may be compared with the mean square within diets $0{\cdot}20161\,(1-0{\cdot}20161)=0{\cdot}16096$. The ratio of these, $0{\cdot}51123/0{\cdot}16096=3{\cdot}18$, may be tested as a variance ratio with 2 and an infinite number of degrees of freedom. Reference to *Table III* shows that as high a value as 3·00 would occur by chance in less than 5 per cent of trials. The differences between the proportions may therefore be judged significant.

The comparison of each pair of proportions is slightly more difficult since each group has a different number of animals and, consequently, each proportion has a different accuracy. Thus, if we wish to test the difference between the proportions of survivals on diets *B* and *C*, the standard error is

$$\sqrt{\left[0{\cdot}16096\left(\frac{1}{41}+\frac{1}{40}\right)\right]}=\pm 0{\cdot}089$$

while the standard error testing the difference between diets *A* and *B* is

$$\sqrt{\left[0{\cdot}16096\left(\frac{1}{43}+\frac{1}{41}\right)\right]}=\pm 0{\cdot}088$$

Where the numbers in the groups are as similar as here there is little point in calculating each standard error separately, and the standard error of the difference between the two smallest groups might be used. It should be noted that when we have shown that a difference exists between the groups, the overall proportion ceases to have any real meaning. Thus the use of a standard error based upon this proportion is not strictly legitimate and can only be regarded as an approximate indication. To overcome this difficulty completely it would be necessary to calculate the combined proportion for each pair of diets. For example, the difference between *A* and *B* should be tested as follows:

Overall proportion of survivals on diets *A* and *B*

$$=(10+12)/(43+41)$$
$$=0{\cdot}26190$$

Mean square $=0{\cdot}26190\;(1-0{\cdot}26190)$
$=0{\cdot}19331$

Standard error of difference between diets *A* and *B*

$$=\sqrt{\left[0{\cdot}19331\left(\frac{1}{43}+\frac{1}{41}\right)\right]}$$
$$=\pm 0{\cdot}096$$

Normally, of course, extra calculations of this type are unnecessary but they may be carried out without undue difficulty.

5.3 *The chi-squared test*—In testing the differences between the proportions observed in several groups, the variance-ratio test is used with a denominator possessing an infinite number of degrees of freedom. Since in this test the second entry in the variance-ratio table is always infinite it is convenient to compile a single entry table employing only the degrees of freedom of the numerator. Furthermore, it is convenient to avoid dividing the numerator by its degrees of freedom. The resulting quantity is known as chi-squared and is denoted by the symbol χ^2. *Table VI* gives the values of chi-squared corresponding to different levels of probability and for different degrees of freedom (of the numerator). The number of degrees of freedom of any chi-squared is usually shown inferior in parentheses after the symbol. Thus $\chi^2_{(3)}$ represents a chi-squared with 3 degrees of freedom.

To demonstrate the use of *Table VI* consider the example in the last section. The value of $\chi^2_{(2)}$ here is 6·35, and *Table VI* shows that the 5 per cent significance level is 5·99. These values are double those used in the variance-ratio test, but the conclusion to be derived is, of course, the same.

From the above remarks it may be seen that $\chi^2_{(m)}$ tests the ratio of a sum of squares to the theoretical mean square and the average value of $\chi^2_{(m)}$ will in consequence be m. Certain important results follow from this observation.

First, since $\chi^2_{(m)}$ is a sum of squares it will tend to be normally distributed for large m (since it will then be determined by the sum of a large number of independent quantities). This provides a possible test for $\chi^2_{(m)}$, which may be tested as a normal variate with mean m and variance $2m$ for large values of m. However, the test indicated at the bottom of *Table VI* is more suitable unless m is very large.

Secondly, since $\chi^2_{(m)}$ is a sum of squares derived, in effect, from the overall comparison of a series of proportions, other χ^2's which may be derived from particular comparisons or groupings of these proportions will be less than this value. This is liable to be important where, for example, we have a $\chi^2_{(3)}$ of, say, 3·0 derived from the comparison of four proportions. Since no particular comparisons of these proportions or grouping of the data could give a $\chi^2_{(1)}$ exceeding this value, no such comparison could be significant.

Lastly, just as we may add two items in the analysis of variance and test them jointly *e.g.* we may test the variation due to rows and columns in a Latin square simultaneously, so we may add two independent χ^2's together and test them simultaneously. This is called the additive property of χ^2. It is not necessary that the χ^2's should have been derived from experiments carried out at the same time for this to be a valid procedure, since, as we know the theoretical mean squares, no problem is raised by their possible inequality. For example, if four experiments carried out at different times to compare the proportions of survivals on two different diets gave values of $\chi^2_{(1)}$ equal to 3·1, 2·3, 1·9 and 3·4, these may be added to give a $\chi^2_{(4)}$ of 10·7. Although none of the individual experiments is significant, the consistency of values greater than unity gives rise to an overall significant value. This provides quite a quick method of combining tests employing the chi-squared test, but it is not always the best method. The addition of χ^2's in this manner does not take account either of the numbers of observations or of the signs of the differences which are tested. Alternative methods to take these into account will be given later.

5.4 *An alternative computational procedure*—It is possible to calculate values of chi-squared by an alternative procedure which is more generally applicable. For this it is first necessary to use all observations of both a positive and a negative nature *i.e.* we include deaths and survivals. These are therefore set out in a table. For example, the results of the experiment analysed in section 5.1 would be presented as follows:

Diet	*Survivals*	*Deaths*	*Total*
A	10	33	43
B	12	29	41
C	3	37	40
Total	25	99	124

The expected numbers in each group are next calculated assuming the proportions of survivals are all equal. Since 25 out of 124 mice have survived, with 43 mice we should expect $(25/124)\times 43 = 8{\cdot}67$ mice to survive. Similarly, with 41 and 40 mice we should expect $(25/124)\times 41 = 8{\cdot}27$ and $(25/124)\times 40 = 8{\cdot}06$ to survive. These fractional values should be used even though they cannot be realized in practice. In general, the expected number in any class can be derived by multiplying together the two marginal totals corresponding to this class and dividing by the total number of observations. Thus, the expected number of deaths on diet *B* is $(41\times 99)/124 = 32{\cdot}73$. The expected values may now be set out:

Diet	*Survivals*	*Deaths*	*Total*
A	8·67	34·33	43·00
B	8·27	32·73	41·00
C	8·06	31·94	40·00
Total	25·00	99·00	124·00

The last steps in the calculation are to subtract each expected value from the corresponding observed value; to square the difference and divide it by the expected values. These quantities are then summed for all groups to give the value of χ^2. If O represents the observed values and E the expected values, this process gives

$$\chi^2 = \Sigma \frac{(O-E)^2}{E}$$

where the summation occurs over all groups. In this example the contributions from each group are as follows:

Diet	*Survivals*	*Deaths*	*Total*
A	0·204	0·052	—
B	1·682	0·425	—
C	3·177	0·802	—
Total	—	—	6·342

The total obtained by this method agrees with that previously obtained apart from a small difference due to rounding off.

It may be seen that this approach is more lengthy but it may be applied to a wider range of problems. These will be discussed in the next few sections.

It is not always necessary to calculate each difference between the observed and expected values and we may, alternatively, calculate

$$\chi^2 = \Sigma \frac{O^2}{E} - N$$

where N is the total number of observations. The use of this formula might be compared with the overall correction of a sum of squares for the mean, while the previous formula, in effect, corrects each term separately.

5.5 *Comparison of several proportions*—So far it has been assumed that it is desired to test the differences between the proportions with a given attribute in several groups. If each group is classified according to several attributes, then we may desire to test whether the proportions of members in the classes differ from group to group. For example, the groups may be classified according to eye colour: brown, grey or blue. The proportions of individuals with each eye colour may be found and we may then want to test whether these sets of proportions are constant over all groups.

The appropriate test is still the χ^2 test, but this should be calculated from the formulae of the last section. The degrees of freedom will, however, be altered. If there are r proportions to be compared between s groups, the number of degrees of freedom is $(r-1)(s-1)$. The reason for this is not very difficult to see. When we have determined $r-1$ proportions in each group, the other is naturally determined. For example, if we know the proportion of survivals, the proportion of deaths is also known. Furthermore, we may make $s-1$ comparisons between s groups (this is the usual number of degrees of freedom). Consequently, there are effectively $(r-1)(s-1)$ comparisons that may be made between the $r-1$ proportions. This determines the number of degrees of freedom. The following example will serve to demonstrate the test of several proportions.

A series of samples was taken of the plant *Calluna* in five different areas and this was classified according to three types: erect, spreading and bushy. The results of this sampling were as follows:

Area	*Erect*	*Spreading*	*Bushy*	*Total*
1	20	45	24	89
2	19	19	26	64
3	55	40	22	117
4	11	23	11	45
5	3	20	10	33
Total	108	147	93	348

It was required to test the differences between the proportions of each type in the different areas so that the first step was to calculate the expected numbers in each group:

Area	*Erect*	*Spreading*	*Bushy*	*Total*
1	27·62	37·59	23·78	88·99
2	19·86	27·03	17·10	63·99
3	36·31	49·42	31·27	117·00
4	13·97	19·01	12·03	45·01
5	10·24	13·94	8·82	33·00
Total	108·00	146·99	93·00	347·99

There are slight differences in the marginal total due to rounding off. Since the lowest number expected in any group is 8·8, the normal approximation is not invalidated. Here, the contributions to χ^2 from the individual groups are as follows:

Area	*Erect*	*Spreading*	*Bushy*	*Total*
1	2·102	1·461	0·002	—
2	0·037	2·386	4·632	—
3	9·620	1·796	2·748	—
4	0·631	0·837	0·088	—
5	5·119	2·634	0·158	—
Total	—	—	—	34·251

The total χ^2 of 34·25 has $(3-1)(5-1)=8$ degrees of freedom and would occur by chance less than once in a thousand times. Inspection of the observed and expected figures shows that this may be attributed largely to the high proportion of erect plants in area *3*. We might, therefore, omit this area and repeat the analysis. The contributions from the different groups are then:

Area	*Erect*	*Spreading*	*Bushy*	*Total*
1	0·086	0·345	0·410	—
2	1·271	3·825	2·037	—
4	0·046	0·224	0·579	—
5	2·759	1·451	0·001	—
Total	—	—	—	13·034

The value of χ^2 has been appreciably reduced, but since it now has $(3-1)(4-1)=6$ degrees of freedom it is still significant at the 5 per cent level. Here the low proportion of spreading plants in area *2* seems to be largely responsible. If this area is excluded, $\chi^2_{(4)}$ testing the difference between areas *1*, *4* and *5* takes the insignificant value of 3·36.

In general, it is a dangerous procedure to remove obviously significant differences until an insignificant value is obtained, since a significant comparison may be ultimately obscured by other insignificant comparisons. Thus it would be necessary to examine likely comparisons for significance. In this example, since the remaining four degrees of freedom total only 3·36, no such comparison could exceed this value *i.e.* could reach significance. It is, therefore, unnecessary to test any comparisons between areas *1*, *4* and *5*.

Two other tests may be carried out on this example. Since area *3* is obviously significantly different from areas *1*, *4* and *5* in the proportion of erect plants, we might test whether the relative proportions of spreading and bushy plants are comparable. Similarly we might test whether the relative proportions of erect and bushy plants in area *2* is different from that in areas *1*, *4* and *5*. To make these tests, we require the 2×2 tables:

Area	*Spreading*	*Bushy*	*Total*
3	40	22	62
1+4+5	88	45	133
Total	128	67	195

Area	*Erect*	*Bushy*	*Total*
2	19	26	45
1+4+5	34	45	79
Total	53	71	124

In both of these tables the percentages are so similar that no formal testing is required. Evidently, the only significant effects are an excess of erect plants in area *3* and a deficit of spreading plants in area *2*.

5.6 *Testing 2×2 tables*—The testing of the difference between two proportions may be conveniently and rapidly carried out using 2×2 tables of the type shown above. In general, if a total of N observations is taken and distributed between the groups as follows:

	First classification		*Total*
Second classification	a	b	$a+b$
	c	d	$c+d$
Total	$a+c$	$b+d$	N

the value of $\chi^2_{(1)}$ testing the difference between the proportions in the two groups is given by

$$\chi^2_{(1)} = \frac{N\,(ad-bc)^2}{(a+c)(b+d)(a+b)(c+d)}$$

This formula may be directly calculated without much difficulty. It is particularly suitable for calculation using logarithms.

The following table gives the deaths and survivals of two groups of mice classified according to the weight gains in the three weeks prior to infection with Salmonella.

Class	*Weight gain* *Low*	*High*	*Total*
Deaths	25	14	39
Survivals	23	30	53
Total	48	44	92

The value of $\chi^2_{(1)}$ for this experiment is

$$\frac{92\,(25\times 30-14\times 23)^2}{48\times 44\times 39\times 53} = \frac{92\times 428^2}{48\times 44\times 39\times 53} = 3{\cdot}86$$

As high a value as this would occur about once in twenty times by pure chance suggesting that the survival rate is higher for animals with higher weight gains.

5·7 *Yates' correction for continuity*—It was pointed out in section 5.1 that in testing the difference between a series of proportions we assume that sufficient observations are taken to ensure that the tested proportions are normally distributed. Otherwise the discrete changes in the values observed will cause the significance of any set of differences to be incorrectly estimated. Where we are testing the difference between two proportions we may improve the approximation by a method due to F. YATES. He has shown that if each observed value in the 2×2 table is altered by $\frac{1}{2}$ to make the observed difference less extreme, the normal approximation has greater validity. Thus, in the above example, we have more deaths than might be expected in the group with low weight gains and the observed number 25 should be altered to 24·5. Similarly, the other classes are altered to give:

Class	*Weight gain* *Low*	*High*	*Total*
Deaths	24·5	14·5	39·0
Survivals	23·5	29·5	53·0
Total	48·0	44·0	92·0

The marginal totals will remain unaltered.

The logic of this correction should be noted. The discrete value 25 effectively corresponds to the range of values 24·5 to 25·5 for continuous variables. The replacement, then, of 25 by 24·5 will ensure that significance cannot arise due to the discrete changes in the proportions *i.e.* due to the limited sample size.

The calculation of $\chi^2_{(1)}$ may then be carried out as before. Here

$$\chi^2_{(1)} = \frac{92\,(24{\cdot}5 \times 29{\cdot}5 - 14{\cdot}5 \times 23{\cdot}5)^2}{48 \times 44 \times 39 \times 53}$$

$$= \frac{92 \times 382^2}{48 \times 44 \times 39 \times 53} = 3{\cdot}08$$

As high a value of $\chi^2_{(1)}$ as this would occur by chance in less than 10 per cent of trials (about 7·9 per cent of trials). The difference between the proportions is still suggestive, but not as large as the previous test indicated. The exact percentage will lie somewhere between the values given by these two methods. It is usually nearer to the percentage obtained when the Yates' correction is employed.

The calculation may be shortened if we subtract half the total number of observations from the quantity $ad-bc$ (sign ignored) before squaring it. This is expressed by the formula

$$\chi^2_{(1)} = \frac{N\,[\,|\,ad-bc\,| - 0{\cdot}5N]^2}{(a+c)(b+d)(a+b)(c+d)}$$

Here therefore we might calculate

$$|\,ad-bc\,| - 0{\cdot}5N = |\,25 \times 30 - 14 \times 23\,| - 0{\cdot}5 \times 92$$
$$= 428 - 46 = 382$$

to give the same value as previously.

This correction is of greatest importance where numbers less than ten are involved, but even for numbers between ten and a hundred it may produce a noticeable effect. Since the increase in the amount of computation is very small, this correction is to be recommended for general application.

5.8 *Testing goodness of fit*—The application of χ^2 in comparing observed with expected values is not restricted to comparing observed proportions. It may also be used to compare an observed proportion with an expected proportion or a series of observed proportions with a series of expected proportions. To consider a simple example, in Scotland in 1947 out of 113·1 thousand live births, 58 thousand were male; suppose we wish to test whether this is consistent with the hypothesis that 51 per cent of live births are male. The observed proportion is 58·0/113·1 = 0·51282 and, if 51 per cent is the true proportion, its standard error is

$$\sqrt{(0{\cdot}51 \times 0{\cdot}49/113{,}100)} = \pm 0{\cdot}001486$$

Since the observed proportion is distributed approximately normally, we may test the difference between the observed and expected values using the normal deviate

$$(0{\cdot}51282-0{\cdot}51000)/0{\cdot}001486=1{\cdot}898$$

As high a deviate as this would occur slightly more than once in twenty times. There is thus some reason for doubting the hypothesis but we cannot completely reject it.

We may carry out the above test using the formula given in section 5.4 for calculating χ^2. Here the analysis proceeds as follows:

Class	*Live births* *Male*	*Female*	*Total*
Observed	58,000	55,100	113,100
Expected	57,681	55,419	113,100
$(O\text{-}E)^2/E$	1·76	1·84	3·60

The expected values are 51 and 49 per cent of the total and the resulting $\chi^2_{(1)}$ of 3·60 would occur slightly more than once in twenty times by pure chance. This value of χ^2 is of course the square of the above normal deviate. It must be noted that if this approach is used, all classes *i.e.* both male and female births, must be included.

Again, the use of χ^2 may be more lengthy if only two classes are involved, but where several classes are involved this method is faster. The application of χ^2 then is demonstrated in the example below.

Figure 4 shows the proportions of different scores obtained by throwing a die. After 240 throws the following frequencies had been observed: one, 43; two, 49; three, 23; four, 41; five, 39; six, 45. If we test whether these frequencies are in agreement with the hypothesis of an equal probability for all numbers, the expected frequency in each class is 40. Consequently

$$\chi^2_{(5)}=\frac{3^2}{40}+\frac{9^2}{40}+\frac{17^2}{40}+\frac{1^2}{40}+\frac{1^2}{40}+\frac{5^2}{40}$$
$$=10{\cdot}15$$

This value is high but it would occur once in ten times by pure chance. The extension of this series of throws to 480 gave frequencies: one, 89; two, 88; three, 62; four, 77; five, 79; six, 85. The expected number in each class is 80 and we now have

$$\chi^2_{(5)}=\frac{9^2}{80}+\frac{8^2}{80}+\frac{18^2}{80}+\frac{3^2}{80}+\frac{1^2}{80}+\frac{5^2}{80}$$
$$=6{\cdot}30$$

This is in closer agreement with the expected value. Evidently, there is no reason for believing the die to be biased.

5.9 *Goodness of fit with estimated constants*—In the examples of the last section, the expected proportions in each group were assumed to be known before the observations were taken. It is only necessary, then, to multiply these proportions by the total number of observations to give the expected numbers in each group.

Sometimes, however, we wish to test whether the observed proportions in each group are in agreement with a general law rather than a set of expected proportions. Thus we may wish to test whether a series of observations in a frequency table is normally distributed. Such general hypotheses usually require the estimation of certain constants before they can be put to the test. For example, to test for normality we might estimate the mean and standard deviation, from which the proportions of observations falling in each grouping interval could be calculated. These calculated proportions would then have to be compared with the observed proportions using the χ^2 test.

The fact that constants have to be estimated from the data will have to be acknowledged in the application of a χ^2 test. Obviously, if sufficient constants are fitted, a very close representation of the data might be obtained. As more constants are fitted so we should expect the value of χ^2 to decrease until, when as many constants as classes are used, a perfect fit is obtained and χ^2 is zero. This fact is acknowledged if χ^2 is reduced by one degree of freedom for every constant determined from the observations. Hence, if one constant is estimated from the data (in addition to the total number of observations) the degrees of freedom will be two less than the number of groups.

This reduction in the degrees of freedom will be strictly valid only if, in estimating the constants, the group frequencies enter linearly. This is in general true since it is very seldom that a complicated function of these frequencies must be taken. However, even where this is not true the reduction in the degrees of freedom is still approximately valid.

As an example we shall consider a set of bacterial colony counts [data of Jones, P. C. T. and Mollison, J. E. *J. gen. Microbiology* 2(1948)54].

For a random distribution of colonies the frequencies with which 0, 1, 2, . . . colonies are observed on a slide can be determined if the mean density of colonies is known*. Thus in testing for randomness the expected frequencies are fitted so as to make the mean density the same as the observed mean density. In consequence the appropriate number of degrees of freedom is two less than the number of groups.

Number of colonies	*Observed frequency*	*Expected frequency*
0	11	14·6
1	37	40·9
2	64	57·2
3	55	53·4
4	37	37·4
5	24	20·9
6 and over	12	15·6
	240	240·0

* This is the Poisson distribution discussed in section 2A.9.

Here the test of goodness of fit is made by

$$\chi^2_{(5)} = \frac{3{\cdot}6^2}{14{\cdot}6} + \frac{3{\cdot}9^2}{40{\cdot}9} + \ . \ . \ . \ + \frac{3{\cdot}6^2}{15{\cdot}6}$$

$$= 3{\cdot}41$$

This value might easily occur by chance so that we conclude that the observed frequencies agree with the hypothesis of a random distribution.

Here again in employing χ^2 we must ensure that the expected frequencies never fall below five. To achieve this it may often be necessary to combine some groups of observations. This has been done in the above example for observations of 6 and over.

SUMMARY OF PP 82 TO 94

The variance-ratio test may be used to determine the significance of the differences between a series of proportions. A modification of the variance ratio known as chi-squared is more suitable for testing the differences between sets of proportions. This may be calculated using the formula

$$\chi^2 = \Sigma \frac{(O-E)^2}{E}$$

where O is the observed number in each group, E is the corresponding expected number in each group, and the summation occurs over all groups. The same formula may be used to test the goodness of fit of a set of calculated values.

A shortened formula has been given to test the difference between two proportions

$$\chi^2_{(1)} = \frac{N(\,|\,ad - bc\,|\, - 0{\cdot}5N)^2}{(a+c)(b+d)(a+b)(c+d)}$$

EXAMPLES

42 In an experiment to test the effects of six different diets, mice on each diet were infected and the numbers of survivals observed. The following results were obtained:

Diet	*A*	*B*	*C*	*D*	*E*	*F*
Survivals	20	43	46	24	24	35
Total No.	25	53	75	27	33	47

Show that the difference between these proportions is significant at the 5 per cent level.

43 The proportions of erect plants of the species *Erica* were observed on grazed and ungrazed plots. Of 59 plants observed on grazed plots, 34 were erect; while of 87 plants observed on ungrazed plots, 70 were erect. Show that such a difference would occur by chance less than once in a hundred times.

44 In a sociological investigation two groups of children were graded *A*, *B* or *C* according to ability. The following numbers were observed:

Grade	*A*	*B*	*C*
Group 1	83	49	30
Group 2	306	99	25

Show that the difference between these groups is highly significant ($\chi^2_{(2)} = 30{\cdot}0$).

45 Using the data of example *2* test whether the proportion of male students with scores exceeding zero is significantly less than the proportion of female students ($\chi^2_{(1)} = 11{\cdot}0$).

This provides a rapid, but wasteful, method of testing the difference between two groups of observations.

46 The following figures give the gradings of a group of children in tests of ability in English and ability in Arithmetic. Use the χ^2 test to determine whether the two abilities are associated.

		Ability in English			*Total*
		A	*B*	*C*	
Ability	*A*	32	12	4	48
in	*B*	5	48	10	63
Arithmetic	*C*	1	2	20	23
Total		38	62	34	134

Here we have a crude method of testing for association between two variables. It should be noted that this test does not recognize the ordering in the classes *A*, *B* and *C*; the same result would be obtained whatever their order.

47 A series of observations on the frequency of occurrence of a mutation in four different areas gave the following results:

Area	*A*	*B*	*C*	*D*
Frequency of mutation	19	10	8	5
Number of observations	484	525	474	527

Show that the difference in frequency between the four areas is significant at the 1 per cent level.

48 Jones and Mollison have published data on the distribution of soil bacterial counts. The following figures give the observed distribution and a distribution fitted after constants measuring the mean density and mean tendency to grouping had been estimated:

Number of bacteria	*Observed frequency*	*Expected frequency*	*Number of bacteria*	*Observed frequency*	*Expected frequency*
0	11	13·0	7	16	16·7
1	17	21·0	8	13	14·1
2	31	24·6	9	17	11·7
3	24	25·4	10	6	9·6
4	29	24·2	11	8	7·8
5	18	22·0	12 and over	31	30·5
6	19	19·4	*Total*	240	240·0

Show that the value $\chi^2_{(10)}$ measuring the goodness of fit is 8·38 and hence conclude that the fitted method of representation is adequate.

EXTENDED DEVELOPMENT

5A.10 *Components of chi-squared*—In the same manner as we may select and test particular comparisons when we are dealing with arithmetic means, so we may select and test particular comparisons of proportions. This can be done by the same method as was used in section 4A.9; the residual mean square being calculated using all groups. This is demonstrated in the following example:

The incidence of strawberry footrot was observed on the four feet of 84 ewes. The numbers of affected feet were: left hind, 65; right hind, 56; left fore, 47; right fore, 43. We may test whether the incidence was equal on all four feet (each observation being assumed independent) either using the differences between observed and expected numbers or directly comparing proportions as in section 5.2. The analysis by the latter method proceeds in the following manner:

Foot	*Number of feet, n*	*Number of affected feet, m*	*Proportion of affected feet, m/n*	m^2/n
Left hind	84	65	0·774	50·2976
Right hind	84	56	0·667	37·3333
Left fore	84	47	0·560	26·2976
Right fore	84	43	0·512	22·0119
Total	336	211	0·6280	135·9404

Overall correction term $=(211)^2/336=132{\cdot}5030$.

$$\chi^2_{(3)}=\frac{135{\cdot}9404-132{\cdot}5030}{0{\cdot}6280\,(1-0{\cdot}6280)}=14{\cdot}71$$

This value of χ^2 is highly significant. Suppose we now wish to compare the hind with the fore feet. The mean difference is

$$\tfrac{1}{2}(0{\cdot}774+0{\cdot}667)-\tfrac{1}{2}(0{\cdot}560+0{\cdot}512)=0{\cdot}1845$$

The standard error of each proportion is

$$\sqrt{[0{\cdot}6280\,(1-0{\cdot}6280)/84]}=0{\cdot}05274$$

and the standard error of this difference is consequently

$$0{\cdot}05274\sqrt{(1/4+1/4+1/4+1/4)}=0{\cdot}05274$$

The normal deviate $0{\cdot}1845/0{\cdot}05274=3{\cdot}4983$ is significant at the 0·1 per cent level. Alternatively $(3{\cdot}4983)^2=12{\cdot}24$ may be tested as a $\chi^2_{(1)}$. Here, since the numbers upon which each group is based are equal, the same result could have been obtained by adding the groups together and calculating χ^2, although this could not be done in general.

Since this comparison accounts for 12·24 of the total $\chi^2_{(3)}$ of 14·71 the remaining $\chi^2_{(2)}$ of 2·47 is not significant. Evidently, the difference between the feet arises roughly from differences between the fore and hind feet.

Other comparisons may be made in the same manner. Thus the difference between the right and left feet is tested by a $\chi^2_{(1)}$ of 2·15, and the difference between the right hind and left fore feet as compared with the left hind and right fore feet by a $\chi^2_{(1)}$ of 0·32. Here, these comparisons are orthogonal since there are equal numbers in each group and consequently the $\chi^2_{(1)}$'s may be added to give the overall $\chi^2_{(3)}$.

Thus:

$$12{\cdot}24+2{\cdot}15+0{\cdot}32=14{\cdot}71$$

This property would not usually occur, but it is often present in genetical applications of χ^2 testing goodness of fit.

In genetical applications χ^2 is usually used to test the agreement between observed frequencies and the frequencies expected under the Mendelian hypothesis. For example, for independent genes, under the Mendelian hypothesis, the double backcross *AaBb* × *aabb* is expected to give equal proportions of the genetical types *AaBb*, *Aabb*, *aaBb* and *aabb*. This can be tested by a $\chi^2_{(3)}$. However, deviations from the expected proportions may occur for any one of three reasons. First, the proportions of genetical types *Aa* and *aa* may not be equal. Secondly, the proportions of genetical types *Bb* and *bb* may not be equal. Lastly, the total proportion of *AaBb* and *aabb* may differ from that of *Aabb* and *aaBb* *i.e.* linkage may exist. We may therefore make three comparisons each of which compares two proportions and each of which may be tested using a $\chi^2_{(1)}$. The sum of these $\chi^2_{(1)}$'s gives the overall $\chi^2_{(3)}$ testing the deviations from the expected numbers.

The position here might be compared with the factorial experiment. The tests of the 1 : 1 segregations for *Aa* : *aa* and *Bb* : *bb* correspond to tests for the main effects, while the linkage test corresponds to the test for interaction.

A similar set of components of χ^2 exists for testing segregations arising from the single backcrosses or the F_2. Again, there are three components testing the expected 3 : 1 or 1 : 1 segregations and the linkage between the genes.

5A.11 *Combination of tests of significance*—It has been observed in section 5.3 that any two χ^2's may be added together to give a third χ^2. This was termed the additive property of χ^2. It provides a rapid method of combining and testing the significance of results from different experiments in which χ^2 is used.

Where χ^2 is not used, tests of significance may be carried out in each experiment and the probabilities of the results arising by chance worked out. Thus, corresponding to each experiment, there is a probability and we have to assess whether this series of probabilities is suggestive of real effects. For example, suppose in testing the effect of a drug in two experiments on different occasions we get probabilities of 0·06 and 0·07 of the results arising by chance. Both of these experiments are highly suggestive and taking them together we should feel reasonably certain of a real effect. However, a definite figure to measure the degree of certainty is desirable, especially where there may be some doubt about the overall significance. Such a figure may be reached using the table of χ^2.

It so happens that the probability P of obtaining any $\chi^2_{(2)}$ as large as a given value χ^2 may be directly calculated (as $\exp-0{\cdot}5\chi^2$). In the same manner the $\chi^2_{(2)}$ corresponding to any probability P can be calculated. This is $-2\log_e P$.

Hence, if we have a series of probabilities we may calculate a corresponding series of $\chi^2_{(2)}$'s by taking minus twice their natural logarithms. The additive property of χ^2 may now be used to give an overall χ^2 for all experiments, which can be tested in the ordinary manner.

If, then, we have two probabilities 0·06 and 0·07 the corresponding values of $\chi^2_{(2)}$ are $-2\log_e 0{\cdot}06 = 5{\cdot}627$ and $-2\log_e 0{\cdot}07 = 5{\cdot}319$. The overall $\chi^2_{(4)}$ testing the two experiments simultaneously is $5{\cdot}627 + 5{\cdot}319 = 10{\cdot}946$ which, from *Table VI*, would occur by chance in about 3 per cent of trials. The same method can generally be employed to combine the probabilities from a series of experiments.

Various points should be noted. This combination of probabilities does not take into account the possible varying accuracies of the experiments from which they are derived; each experiment is given the same weight. In addition, the test does not take into account the signs of the differences tested, thus two experiments with differences in opposite directions may be as significant as two experiments with differences in the same direction. There may, therefore, often be more suitable methods of combining experimental results. The above method is more comprehensive but, for this reason, usually not the best possible.

If we test the difference between two proportions in several experiments, we may not be able to use the combined figures if there is a tendency for these proportions to vary from experiment to experiment. Thus, if we are comparing survivals on two different diets, in one experiment a high percentage of animals may die on both diets, while in a second experiment a low proportion may die. The results from these two experiments cannot be added together, but the $\chi^2_{(1)}$'s from the two experiments might be added. If, however, we wish to take account of the signs of the observed differences, we may calculate the normal deviates *i.e.* square roots of $\chi^2_{(1)}$, testing the differences between the proportions. If these are added and divided by the square root of the number of differences which are tested, the resulting value is still a normal deviate and may therefore be tested. It may alternatively, of course, be squared and tested as a $\chi^2_{(1)}$. The following example brings out these points:

In an experiment to compare the mortality of animals kept under different conditions, the following results were obtained:

Males

Method	*I*	*II*	*Total*
Survivals	13	8	21
Deaths	3	10	13
Total	16	18	34

$$\chi^2_{(1)} = \frac{34\,(89)^2}{16\times18\times21\times13} = 3{\cdot}4253$$

Females

Method	*I*	*II*	*Total*
Survivals	7	1	8
Deaths	12	10	22
Total	19	11	30

$$\chi^2_{(1)} = \frac{30\,(43)^2}{19\times11\times22\times8} = 1{\cdot}5080$$

Neither of these $\chi^2_{(1)}$'s reaches significance, nor does the total $\chi^2_{(2)}$ of 4·9333. However, the differences are in the same direction and the results for the females lend strength to those for the males. There is obviously a difference in the death rates for the males and females so that the figures cannot be combined directly. If, however, we calculate the normal deviates for these experiments, we get $\sqrt{(3{\cdot}4253)} = 1{\cdot}8508$, and $\sqrt{(1{\cdot}5080)} = 1{\cdot}2280$. The value of $\chi^2_{(1)}$ resulting from the combination of these deviates is $(1{\cdot}8508 + 1{\cdot}2280)^2/2 = 4{\cdot}74$, which is significant at the 5 per cent level.

This form of combination still does not take into account the differing accuracies of the separate experiments but it obviously produces a definite improvement in sensitivity. It might be noted that this improvement arises because we take note of the signs of the observed differences.

In testing the difference between two proportions by the χ^2 test, it may sometimes be necessary to consider only differences of a given sign. It may be argued that if a difference exists between the groups it could only occur in one direction and that differences in the opposite direction must be ascribed to chance. For example, we might state that one set of conditions cannot be less favourable to survival than another but that it may be more favourable. Here, since we are interested in deviations in one direction only, probabilities calculated using χ^2 must be halved. We are then said to be using a one-tailed test.

5A.12 *Exact testing of 2 × 2 tables*—The use of the χ^2 test is dependent upon the assumption of normality and, indirectly, upon a sufficiency of observations in each group. If there are not sufficient numbers in each group, then the approximation will break down. Use of the Yates' correction improves this approximation, but even with this adjustment χ^2 cannot be used when the expected numbers are low. On such occasions it is necessary to carry out an exact test of significance. R. A. FISHER has shown how this may be done for the 2 × 2 table. His argument proceeds as follows.

We consider for the same marginal totals the proportion of occasions on which the observed frequencies would occur. Let the observed frequencies be as previously:

	First classification		*Total*
Second	a	b	$a+b$
classification	c	d	$c+d$
Total	$a+c$	$b+d$	N

Then the number of ways in which the total N observations may be distributed with $a+b$ occurring in the first row is $N!/[(a+b)!\,(c+d)!]$ and the number of ways in which they might be distributed for $a+c$ observations to fall in the first column is $N!/[(a+c)!\,(b+d)!]$. Thus there are

$$\frac{N!}{(a+b)!\,(c+d)!} \times \frac{N!}{(a+c)!\,(b+d)!} = \frac{(N!)^2}{(a+c)!\,(b+d)!\,(a+b)!\,(c+d)!}$$

ways in which the N observations may be distributed to have the same marginal totals as have been observed. Of these, only $N!/[a!\,b!\,c!\,d!]$ ways will have the same group totals as have been observed. The proportion of occasions on which the observed frequencies would occur is thus

$$\frac{N!}{a!\,b!\,c!\,d!} \div \frac{(N!)^2}{(a+b)!\,(c+d)!\,(a+c)!\,(b+d)!} = \frac{(a+b)!\,(c+d)!\,(a+c)!\,(b+d)!}{N!\,a!\,b!\,c!\,d!}$$

This gives the probability of obtaining the observed frequencies given the marginal totals $a+b$, $c+d$, $a+c$ and $b+d$. We require, in general, the probability of getting at least as extreme a set of frequencies as those observed, so that a series of such probabilities has to be summed for more extreme differences. Thus in testing the female survivals in the last section we should require the probabilities of observing:

7	1	8
12	10	22
19	11	30

and a more extreme difference with the same marginal totals:

8	0	8
11	11	22
19	11	30

These are given by

$$\frac{8!\,22!\,19!\,11!}{30!\,7!\,1!\,12!\,10!} = 0{\cdot}0963$$

$$\frac{8!\,22!\,19!\,11!}{30!\,8!\,0!\,11!\,11!} = 0{\cdot}0131$$

The probability of getting as extreme a difference (of the same sign) as that observed is thus $0{\cdot}0963 + 0{\cdot}0131 = 0{\cdot}1094$. However since this tests only differences in one direction, the probability must be doubled to give a value comparable with that obtained using the χ^2 test in the last section. The resulting value 0·2188 is in fairly close agreement with the probability, 0·2196, which may be derived from the $\chi^2_{(1)}$ of that section. Evidently, for this example, Yates' correction is quite adequate.

This approach involves more calculation than the χ^2 test especially if the number of terms to be evaluated is large so that, in general, the latter approach should be applied if possible. For occasions when this is not possible, D. J. FINNEY [*Biometrika* 35 (1948) 145] has tabulated the probabilities for different numbers of observations in the groups of the 2×2 table. The scope of this tabulation is limited to fairly small numbers, but it is precisely here that it is of most use.

SUMMARY OF PP 96 TO 101

It has been shown how particular components of chi-squared may be tested and how, on certain occasions, chi-squared may be partitioned.

The use of chi-squared for combining tests of significance has been given and particular methods of combining tests which employ chi-squared have been demonstrated.

Finally, the exact method of testing the difference between two proportions has been shown and compared with the test employing Yates' correction.

EXAMPLES

49 The tests of significance in a series of five experiments gave probabilities of 0·22, 0·47, 0·07, 0·31, and 0·12. Show that the overall significance of these experiments can be tested using $\chi^2_{(10)} = 16{\cdot}44$. Hence conclude that such a series would occur more frequently than once in twenty times but less frequently than once in ten times by pure chance.

50 Test the following 2×2 tables and combine your results to give an overall test of significance:

13	6	19
7	14	21
20	20	40

16	13	29
4	7	11
20	20	40

9	3	12
11	17	28
20	20	40

(The individual $\chi^2_{(1)}$'s are 3·61, 0·50 and 2·99. The overall $\chi^2_{(1)}$ is 6·27.)

51 Show that for the male animals of the example of section 5A.11 the exact test of significance gives the probability

$$2\,(0{\cdot}02641 + 0{\cdot}00412 + 0{\cdot}00032 + 0{\cdot}00001) = 0{\cdot}0617$$

(This might be compared with the value 0·0644 obtained using Yates' correction.)

6

INTERRELATIONS OF SETS OF MEASUREMENTS

6.1 *Associated measurements*—So far the problems encountered in presenting and comparing groups of similar measurements have been considered. This does not, however, cover all the types of problems which are commonly encountered. It is often necessary to take several different measurements at the same time to determine whether they are related or not and, possibly, to determine the form of relationship between the variables. For example, measurements may be taken of height and weight which it is desired to relate, or a series of observations may be studied to determine how they change with time. The association of measurements raises two main problems which are encountered in all fields of inquiry:

1 Is there any association between the observed variables and, if so, to what extent will a knowledge of one variable allow determination of the other?

2 What is the form of the association between the observations and how may any one set be estimated from the other?

Thus in the first instance it is necessary to find out whether the observed variables are related *i.e.* whether changes in one can be accounted for by changes in the other. Often, however, there will be no doubt about the existence of a relationship between the variables. For example, there is little doubt of an association between weight and height. On such occasions we shall want to know the extent and form of the association.

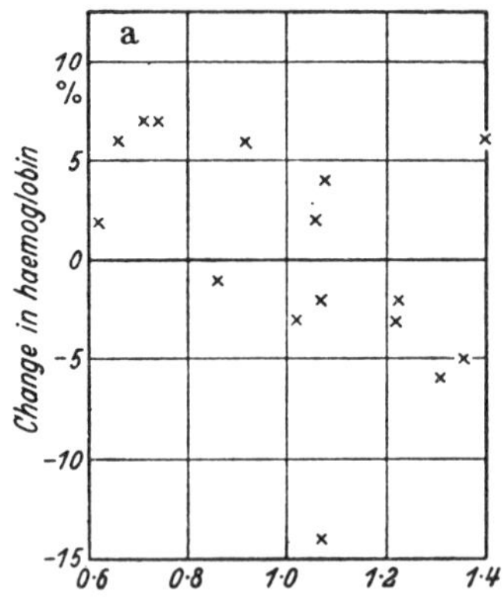

Figure 21a. Scatter diagram of change in haemoglobin percentage of sheep plotted against change in weight

In this chapter, the methods appropriate for testing and determining the association between two sets of measurements will be considered.

6.2 *Diagrammatic presentation*—An initial investigation of the association between two sets of measurements may most easily be carried out using diagrams. By this means it is possible to judge visually whether any association exists, and to assess the form and extent of the association and any irregularities in the observations. Such diagrams do not make statistical tests and methods unnecessary, but they will often facilitate an approach to the data and indicate methods of analysis.

Figure 21 gives two examples of scatter diagrams from experiments in which there is some doubt about the associations between the variables. In the first experiment there is a suggestion that the observed changes in sheep haemoglobin might be related to the changes in weight; while in the second experiment there is a strong indication that the percentage sand is decreasing with depth. In both of these examples it would be necessary to test whether these apparent associations could have arisen by chance. It should be noted that in the first experiment neither variable is subject to control, while in the second experiment the depth at which the measurements are taken may be controlled. We may or may not be able to control one of the variables, but the methods to be considered later will be generally applicable.

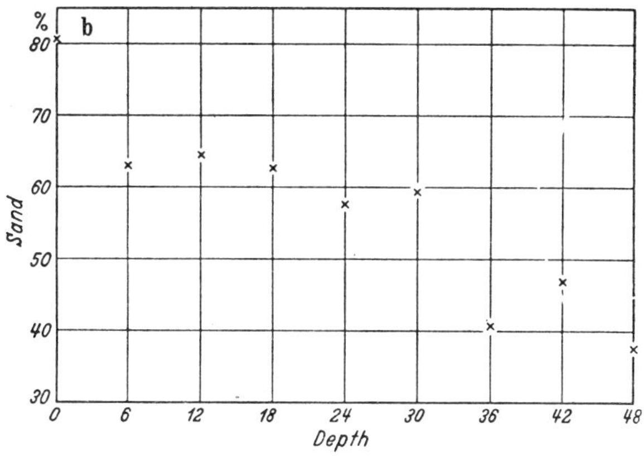

Figure 21b. Plot of percentage sand (by volume) in soil at different depths

Figures 22 and *23* give three examples of measurements which might be expected to be associated. Thus we should expect a large loss of water during drying to be associated with a high specific gravity, a large number of cows to be associated with a high milk production and the heights of babies at birth and one year to be associated. The interest, here, lies in the extent of the association. This may be measured by the extent to which changes in one variable can be accounted for by changes in the other. If the variability in one set of observations can be completely accounted for by differences in the other set of observations, the two variables are said to be perfectly associated. Otherwise the association is partial and may be measured by the proportion of the variability in one set of observations that may be accounted for by changes in the other variable. For example, in *Figure 23*, it appears that

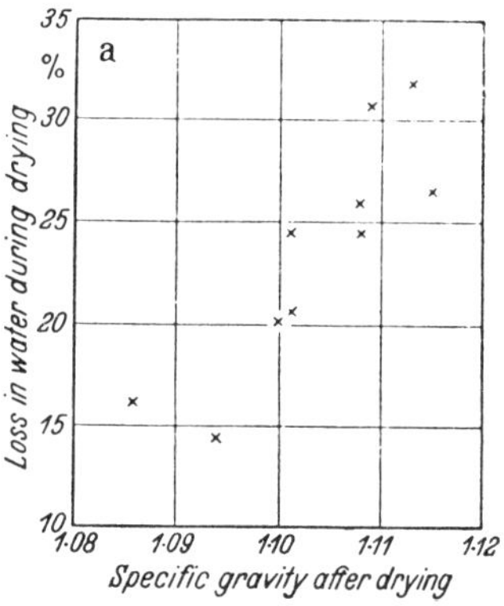

Figure 22a. Percentage loss of water in haddocks during drying plotted against specific gravity after drying

a knowledge of height at birth allows us to predict the height at one year, but not completely. If the height at birth is unknown the height at one year ranges between 66 and 81 cm *i.e.* a range of 15 cm, but a knowledge of the height at birth reduces this range by about one half. For instance, for babies of initial height about 47 to 48 cm the measurements

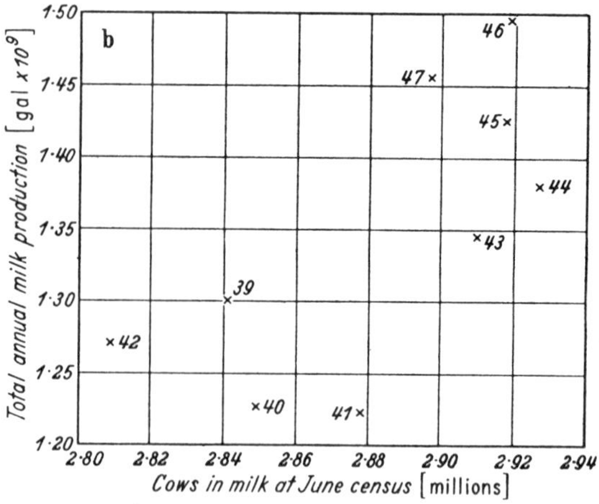

Figure 22b. Annual milk production plotted against cows in milk at June census

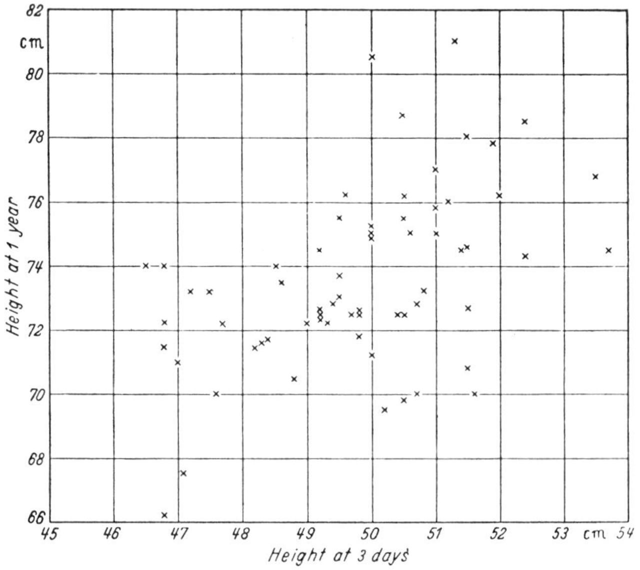

Figure 23. Scatter diagram of heights of male babies at one year against heights at three days

range between 66 and 74 cm, while for initial measurements about 49 to 50 cm the corresponding range of measurements at one year is 71 to 77 cm.

This idea of the extent of the association is important since we often wish to know what use one variable might be in predicting or determining

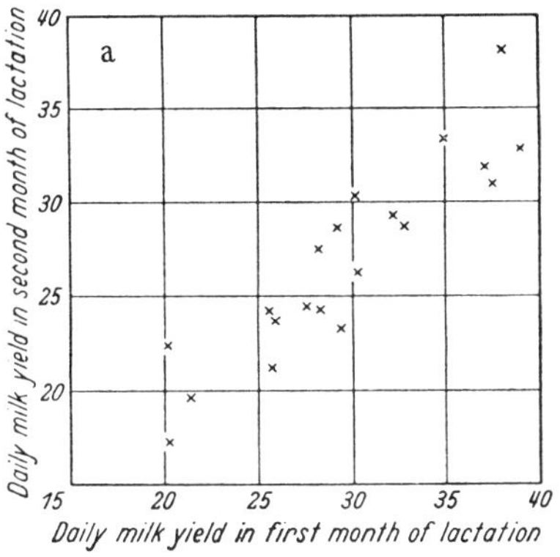

Figure 24a. Plot of milk yields of cows in first and second months of lactation

another. We shall see later that this may be determined using the above concepts.

The association between the variables will often be obvious. The problem then is how to estimate the form of the association. Thus, for the measurements shown in *Figure 24* there is little doubt that the milk yield in the second month of lactation is closely related to that in the first month or that the size of population is closely related to the time of observation. Here the forms of the associations are required. If these can be determined it will be possible to estimate the milk yield in the second month of lactation from the yield in the first month or, possibly, to estimate the future trend of population.

Frequently, measurements which are easily made are used to determine measurements which are more difficult and expensive. Thus, in *Figure 22a*, if specific gravity after drying could have been used to predict the percentage loss of water accurately, this latter, more difficult, measurement would have been unnecessary. In the same manner, if we can predict the life of, say, a radio valve without having to burn it out, we may save a great deal of time and expense.

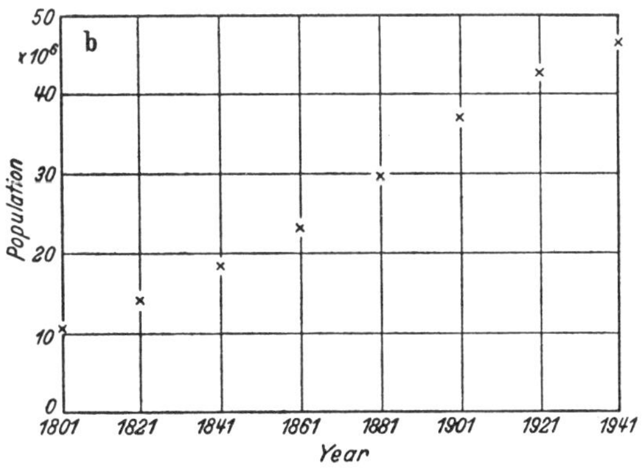

Figure 24b. Changes in the population of Great Britain

There are many other reasons why we might want to find the form of relationship between two variables. It will be shown in Chapter 7 how such relationships can be used to give considerable improvements in experimental accuracy. Before this, however, the problems raised above must be considered.

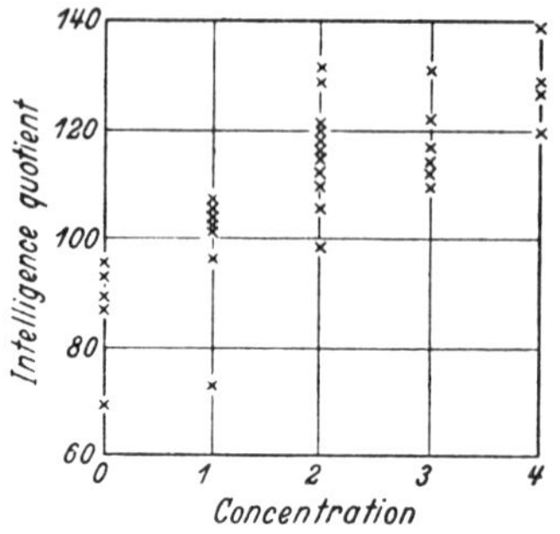

Figure 25. Plot of intelligence quotient against concentration

6.3 *General test for association*—The analysis of variance may be used to provide a general test of association between two variables. If corresponding to each value of one variable there are several measurements of the other variable, the variation due to the former variable may be estimated using the analysis of variance. Thus, in *Figure 25*, degree of concentration has been measured on a five point scale, and an analysis of the variation in intelligence may be carried out as follows:

	D.f.	*S.s.*	*M.s.*	*V.r.*
Between degrees of concentration	4	5,858	1,464·5	15·4
Within degrees of concentration	28	2,672	95·4	
Total	32	8,530	266.6	

The mean intelligence for different degrees of concentration varies significantly *i.e.* there is a significant association between concentration and intelligence. The association between these two variables may be gauged in two ways:

1 By the proportion of the total variation in one variable that can be accounted for by variations in the other. In this example this is 5,858/8,530=0·6868. The square root of this value, here 0·829, is called the correlation ratio and is usually denoted by η. The use of this quantity is very limited since it depends upon the numbers of degrees of freedom in its numerator and denominator.

2 By a comparison of the standard deviation or mean square for the total with the within-group standard deviation or mean square. These indicate the variations in one set of measurements according to whether the other measurement is fixed or not. Thus, in the above example, the standard deviation of the intelligence quotients is $\sqrt{(266{\cdot}6)} = \pm 16{\cdot}3$, but given the degree of concentration this is reduced to $\sqrt{(95{\cdot}4)} = \pm 9{\cdot}8$. If the degree of concentration is known it is easier to predict the intelligence quotient.

This approach takes no account of the ordering of the groups since it tests a relationship between the variables of the general type. For testing relationships of a special kind, more sensitive tests can be devised. Further, this method tests whether knowledge of one variable helps to predict the other but not *vice versa*. This distinction may be of importance where a complicated relationship is to be tested. Thus, in *Figure 26*, the death rate can be predicted for a given age but not *vice versa*.

In the above application it is necessary for several measurements of one variable to correspond to each measurement of the other. In practice this does not often occur but we may still split the data into several groups and test the differences between them. Hence, in testing the data of *Figure 23*, we may group the initial heights in 1 cm groups *e.g.* 46-, 47- *etc* and carry out the following analysis, which is significant at the 1 per cent level:

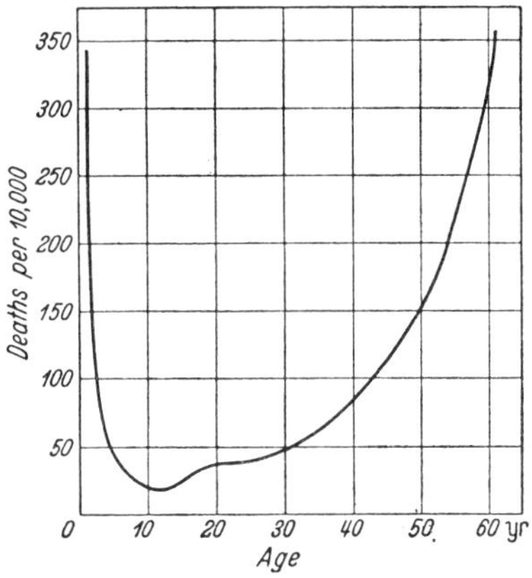

Figure 26. Male death rates for England and Wales in 1911

	D.f.	*S.s.*	*M.s.*	*V.r.*
Between initial heights	7	137·35	19·62	3·11
Within initial heights	57	359·05	6·30	
Total	64	496·40	7·76	

Whereas there is still a slight variation in initial height within the groups this is very small and may be neglected. (This grouping accounts for 98 per cent of the total variation in the initial heights.) Here the correlation ratio is 0·526 and a knowledge of the initial height reduces the standard deviation from $\surd(7{\cdot}76) = \pm 2{\cdot}79$ cm to $\surd(6{\cdot}30) = \pm 2{\cdot}51$ cm.

In general, we are more interested in testing for association of a particular type so that henceforth we shall consider special types of association.

6.4 *Measures of joint variation*—If there are two sets of observations and it is required to estimate their variation, then the mean squared deviation or variance of each set is calculated. For example, if the sets are 3, 4, 7, 2, 4 and 5, 8, 12, 4, 6, the means are 4 and 7 and the deviations from the means are −1, 0, 3, −2, 0 and −2, 1, 5, −3, −1. The estimated variances are then

$$(1^2+0^2+3^2+2^2+0^2)/4=3{\cdot}5$$

and

$$(2^2+1^2+5^2+3^2+1^2)/4=10{\cdot}0$$

Suppose now the mean product of the deviations of two sets of observations taken in pairs is considered. For the above sets this would be

$$[-1\times(-2)+0\times1+3\times5-2\times(-3)+0\times(-1)]/4=5{\cdot}75$$

where, as previously, the sum of products is divided by one less than the total number of observations. A scatter diagram plotting the two sets of deviations against one another is given in *Figure 27*.

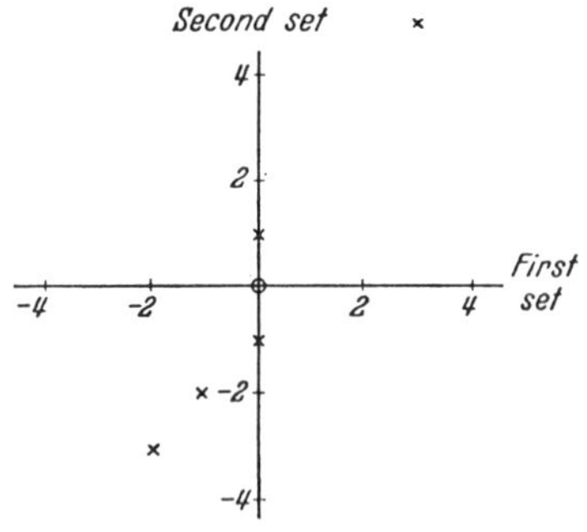

Figure 27. Scatter diagram of five pairs of deviations

If both deviations are positive or negative, then their product will be positive, but if one deviation is positive and the other negative, their product will be negative. Thus, according to the quadrant of the scatter diagram of the deviations in which the point corresponding to each pair of observations falls, it will contribute positively or negatively to the mean product. Points falling in the upper right or bottom left corner will contribute positively, while points falling in the upper left or lower right will contribute negatively. The signs of these contributions are indicated in *Figure 28*.

Now, if two series are unrelated, the scatter diagram will have, on average, about the same number of points in each quadrant. In consequence the mean product of the deviations will be small or zero. But if there is an excess of points in the upper right and lower left quadrants, as in *Figure 27*, the mean product will be positive. Here, an increase in one variable is associated with an increase in the other and they are said to be positively associated. Correspondingly, when a decrease in one variable is associated with an increase in the other, there is an excess of points in the upper left and bottom right quadrants and the mean product is negative. The variables are then said to be negatively associated.

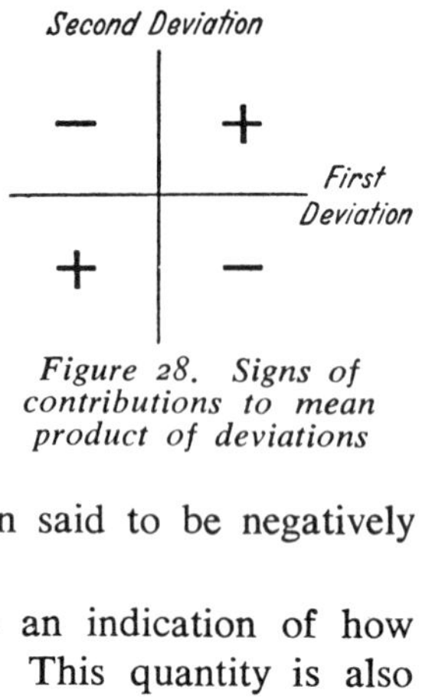

Figure 28. Signs of contributions to mean product of deviations

The mean product of the deviations thus gives an indication of how and to what extent two variables are associated. This quantity is also called the covariance of the two sets of observations. It may be expressed mathematically as follows. If $x_1, x_2, \ldots x_n$ and $y_1, y_2, \ldots y_n$ are two sets of n corresponding observations and their arithmetic means are $\bar{x}$ and $\bar{y}$, then the covariance of the two sets of observations is

$$\frac{1}{n-1}[(x_1-\bar{x})(y_1-\bar{y})+(x_2-\bar{x})(y_2-\bar{y})+\ldots+(x_n-\bar{x})(y_n-\bar{y})]$$

$$=\frac{\Sigma(x-\bar{x})(y-\bar{y})}{n-1}$$

The calculation of this quantity may be simplified by a rule similar to that given in section 1.10 for the calculation of sums of squares of deviations: the sum of the products of deviations of two sets of observations from their means is equal to the sum of products of deviations from any convenient values less the products of the sums of these latter deviations divided by the number of observations. If a and b are any two values this rule is expressed mathematically by the equation

$$\Sigma(x-\bar{x})(y-\bar{y})=\Sigma(x-a)(y-b)-\frac{[\Sigma(x-a)][\Sigma(y-b)]}{n}$$

In particular, if we choose a and b to be zero we get

$$\Sigma(x-\bar{x})(y-\bar{y})=\Sigma xy-\frac{(\Sigma x)(\Sigma y)}{n}$$

These formulae are exactly the same as those used for sums of squares except that wherever a square occurred previously it is now replaced by a product. The term $[\Sigma(x-a)][\Sigma(y-b)]/n$ which corrects the sums of products about arbitrary values a and b to give sums of products about the mean may be called a correction term for the products.

It should be noted that a covariance is most likely to be useful where a straight line relationship exists between the two variables. Otherwise, for relationships of the type shown in *Figure 26*, points will fall into each quadrant because of the nature of the relationship.

As an example of the calculation of the covariance of two series of observations, consider the observations plotted in *Figure 21b*:

										Total
Depth in	0	6	12	18	24	30	36	42	48	216
% Sand	80·6	63·0	64·3	62·5	57·5	59·2	40·8	46·9	37·6	512·4

The sum of products of corresponding observations is

$$0\times80{\cdot}6+6\times63{\cdot}0+\ldots+48\times37{\cdot}6=10{,}674{\cdot}0$$

so that the sum of products of deviations is

$$10{,}674{\cdot}0-\frac{216\times512{\cdot}4}{9}=-1{,}623{\cdot}6$$

The covariance is thus $-1{,}623{\cdot}6/8=-202{\cdot}95$. This value is negative indicating a decrease in percentage of sand with increasing depth. The same result could be obtained if, say, 24 in is subtracted from depth and 60 from the percentage to give:

										Total
Depth − 24	−24	−18	−12	−6	0	6	12	18	24	0
% Sand − 60	20·6	3·0	4·3	2·5	−2·5	−0·8	−19·2	−13·1	−22·4	−27·6

The sum of products here is

$$-24\times 20{\cdot}6-18\times 3{\cdot}0-\ldots-24\times 22{\cdot}4=-1{,}623{\cdot}6$$

as before, since the correction term is zero.

The chief failing of the use of covariance to indicate the association between sets of variables is that it is dependent upon the scale or scatter of the measurements. Thus, by measuring depth in feet instead of inches, we may reduce the covariance by a factor of twelve. This failing may be overcome if the covariance is divided by the standard deviations of both sets of observations. The resulting quantity, which is then independent of any units, is called the correlation coefficient.

In the first example of this section the correlation coefficient is $5{\cdot}75/\surd(3{\cdot}5\times 10{\cdot}0)=0{\cdot}972$. In the above example the variances of depths and percentage of sand are 270·0 and 177·8, so that the correlation coefficient is $-202{\cdot}95/\surd(270{\cdot}0\times 177{\cdot}8)=-0{\cdot}926$.

The correlation coefficient may be expressed mathematically as

$$\frac{\Sigma(x-\bar{x})(y-\bar{y})/(n-1)}{\surd[\{\Sigma(x-\bar{x})^2/(n-1)\}\{\Sigma(y-\bar{y})^2/(n-1)\}]}=\frac{\Sigma(x-\bar{x})(y-\bar{y})}{\surd[\{\Sigma(x-\bar{x})^2\}\{\Sigma(y-\bar{y})^2\}]}$$

and is commonly denoted by r. The true value of this coefficient, which we are attempting to estimate, is denoted by the corresponding Greek letter ρ. In the latter of the above formulae for the correlation coefficient it is unnecessary to calculate the mean squares and products; instead the total sums of squares and products are used directly.

The correlation coefficient, like the covariance, measures the association between two variables, but a fuller consideration of its meaning and significance must wait until section 6.7.

6.5 *Fitting straight lines*—Consider now the most common form of relationship: the straight line. Here for each unit increase in one variable, say x, there is a corresponding increase or decrease β in the other variable. If x_0, y_0 is any particular pair of observations and x, y is a general pair of observations then, corresponding to the change $x-x_0$ in the first variable, there is a change $y-y_0$ in the other variable. Since this latter is β times the former we get

$$y-y_0=\beta(x-x_0)$$

This is an equation for a straight line.

In general, however, the variation in y will not be completely accounted for by changes in x and the series of points will not lie exactly on a straight line. An extra term is then required in this equation to indicate the variation of y about the line. Thus the equation indicating the dependence of y on x may be written

$$y-y_0=\beta(x-x_0)+e$$

where e is an extra term giving the portion of y which cannot be accounted for by variations in x. Hence, if the first variable x is known, the corresponding value of y on the straight line can be estimated, but owing to the extra variation, the observed values will not fall exactly on the line. For example, we may find the mean weight for each height, but the observed values will vary about this mean owing to the extra variation in weight that cannot be accounted for by height.

In specifying the dependence of y on x it is therefore necessary to estimate the equation of the straight line and the variation about it. The line will be chosen so that the residual variation is as small as possible. The sum of squares of the deviations about the line is

$$\Sigma e^2=\Sigma[(y-y_0)-\beta(x-x_0)]^2$$

and the unknown quantities must be chosen to make this a minimum. In consequence it may be shown (see section 6A.8) that:

1 If x_0 is chosen equal to the arithmetic mean $\bar{x}$, the best value for y_0 is the arithmetic mean $\bar{y}$.

2 The best estimate of β is

$$b=\frac{\Sigma(x-\bar{x})(y-\bar{y})}{\Sigma(x-\bar{x})^2}=\frac{\text{covariance of } x \text{ and } y}{\text{variance of } x}$$

3 The variance of b is

$$\frac{\sigma^2}{\Sigma(x-\bar{x})^2}$$

where σ^2 measures the unaccountable variation in y.

From *3* it is seen that if there is no association between x and y the mean value of b^2 is $\sigma^2/[\Sigma(x-\bar{x})^2]$. Thus the mean value of

$$b^2\Sigma(x-\bar{x})^2=\frac{[\Sigma(x-\bar{x})(y-\bar{y})]^2}{\Sigma(x-\bar{x})^2}$$

is σ^2 and this represents the sum of squares in the analysis of variance for y which might be attributed to x. The analysis of variance for y might then be set out as follows:

	D.f.	*S.s.*
Variation ascribable to x	1	$\frac{[\Sigma (x-\bar{x})(y-\bar{y})]^2}{\Sigma (x-\bar{x})^2}$
Residual variation	$n-2$	$\Sigma (y-\bar{y})^2 - \frac{[\Sigma (x-\bar{x})(y-\bar{y})]^2}{\Sigma (x-\bar{x})^2}$
Total	$n-1$	$\Sigma (y-\bar{y})^2$

The significance of the regression may then be tested either by this analysis of variance or by using the residual mean square from this analysis to attach a standard error to b.

If the soil measurements plotted in *Figure 21* are used to estimate the dependence of percentage of sand on depth, then:

$$\bar{x}=24 \text{ in} \qquad \bar{y}=56{\cdot}93 \text{ per cent}$$

$$\Sigma (x-\bar{x})^2=2{,}160{\cdot}00 \qquad \Sigma (x-\bar{x})(y-\bar{y})=-1{,}623{\cdot}60 \qquad \Sigma (y-\bar{y})^2=1{,}422{\cdot}36$$

The estimated slope of the regression line is thus

$$b=-1{,}623{\cdot}60/2{,}160{\cdot}00=-0{\cdot}75167$$

and the fitted line is

$$y-56{\cdot}93=-0{\cdot}75167\,(x-24)$$

This line is shown in *Figure 29*.

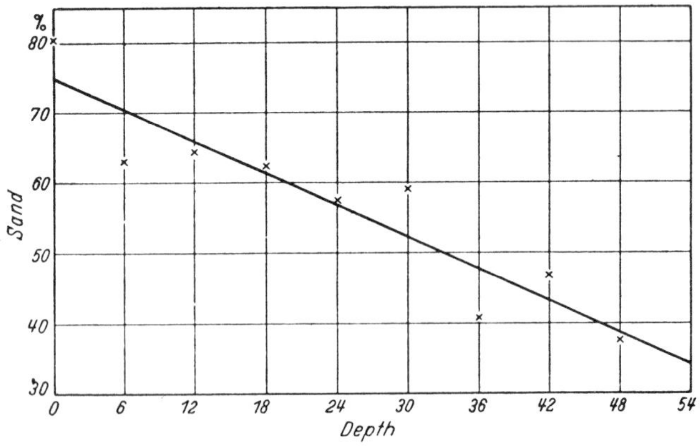

Figure 29 Graph of fitted straight line

If we wish to test the significance of the association, we may calculate the analysis of variance:

	D.f.	*S.s.*	*M.s.*	*V.r.*
Variation ascribable to depth	1	1,220·41	1,220·41	42·3
Residual variation	7	201·95	28·85	
Total	8	1,422·36		

The variance ratio is highly significant. The mean square variation about the line is 28·85 giving a standard deviation of about 5·37 per cent . This roughly agrees with *Figure 29* in which three out of nine points deviate from the line by more than 5·37 per cent, but none deviates by more than 10·74 per cent.

The estimated variance of b is thus 28·85/2,160·00=0·01336 and its standard error is consequently 0·11558. In order to set limits for the value of β, the deviate t with 7 degrees of freedom must be used. Thus, with 99 per cent certainty, β lies between

$$-0{\cdot}75167-3{\cdot}50\times 0{\cdot}11558=-1{\cdot}156$$

and

$$-0{\cdot}75167+3{\cdot}50\times 0.11558=-0{\cdot}0347$$

The equation giving the dependence of y on x is called the regression equation of y on x. Correspondingly, the coefficient b is called the regression coefficient of y on x. In the above example, the percentage of sand y is called the dependent variable and the depth x is called the independent variable.

6.6 *Regression lines*—In giving the equation of a regression line it is necessary to state which is the dependent and which the independent variable. A regression equation shows the manner in which one variable is dependent upon the other, but not *vice versa*. If therefore it is required to predict the depth from the percentage of sand in the same example it would be necessary to use another regression equation

$$x-24=-\frac{1{,}623{\cdot}60}{1{,}422{\cdot}36}(y-56{\cdot}93)$$
$$=-1{\cdot}14148\ (y-56{\cdot}93)$$

The difference between the two regression equations may be appreciated best by a consideration of particular instances where they obviously differ. For example, the mean weight of men 6 ft 3 in tall may be 200 lb, but the mean height of men weighing 200 lb is certainly less than 6 ft 3 in. Another example is provided by *Figure 23*: the mean height at one year of babies of height 47 cm or less at birth is about 70 cm, but the mean height at birth of babies of height 70 cm at one year is about 49 cm. A further example is provided in *Figure 21a*.

In general, the difference between the two regression lines will be greatest when the association between the variables is small. For perfect straight line association the regression lines will of course coincide.

It might now be asked which regression line presents a truer picture of the data. The answer to this is that both regression lines give the most suitable equations for determining the dependent variable from the independent variable, but that neither necessarily gives a true representation of the interrelation of the variables. Thus, for example, neither weight nor height should be considered as causing changes in the other. It is not the change in height which causes the change in weight nor *vice versa;* both weight and height are simultaneously affected by variables on which both depend. Consequently, while one of these variables may be used to predict the other, the regression equation does not represent a natural relationship.

It may, however, happen that one variable is directly dependent upon a second variable and that the whole of the second variable influences the first variable, but is not influenced by it. Here the regression equation of the first variable upon the second will represent the natural relationship between the variables. For example, if for the data in *Figure 29* the depth has been accurately measured, the regression line will represent the effect of depth on percentage of sand. If, however, the depth is not accurately measured then the errors in the measurement of depth will be reflected in the percentage of sand and the regression line will no longer represent the true relationship.

Usually it is required to predict one variable given the other, or to see how much of the variation in one set of measurements can be accounted for by changes in another set. For these purposes the regression equations are needed. It is very seldom that the real form of relationship is required since its use is largely restricted to theoretical problems.

If the regression equations are to be used for estimating the values of one variable from measurements of the other it is necessary to know the errors of the estimates. These can be found using the variances of the estimated mean and regression coefficient. The estimated value Y for any value X is $\bar{y}+b(X-\bar{x})$. The variance of b is $\sigma^2/\Sigma(x-\bar{x})^2$ and therefore the variance of $b(X-\bar{x})$ is $\sigma^2(X-\bar{x})^2/\Sigma(x-\bar{x})^2$. Using the theorem of section 3.5 we thus find the variance of $\bar{y}+b(X-\bar{x})$ as

$$\frac{\sigma^2}{n}+\frac{\sigma^2(X-\bar{x})^2}{\Sigma(x-\bar{x})^2}=\sigma^2\left[\frac{1}{n}+\frac{(X-\bar{x})^2}{\Sigma(x-\bar{x})^2}\right]$$

If the estimated value is to be compared with a further observation or mean, then the above variance must be added to that of the further observation.

In *Figure 29* the fitted line gives an expected percentage of sand of 66·0 at 12 in depth. The variance of this estimate is

$$28{\cdot}85\left[\frac{1}{9}+\frac{(12-24)^2}{2{,}160}\right]=5{\cdot}13$$

The standard error is $\sqrt{(5{\cdot}13)}=\pm 2{\cdot}26$ with 7 degrees of freedom.

A further sample at the same depth taken later contained 58·1 per cent of sand. The difference between this sample and the estimate is 7·9 and its standard error is $\sqrt{(28{\cdot}85+5{\cdot}13)}=\pm 5{\cdot}8$. Evidently the extra sample lies within the range of sampling variation.

The above approach may be employed within the range of the observations used to calculate the regression line, but it is dangerous to attempt to estimate values falling outside this range. If it is certain that the same straight line relationship will hold outside the range then standard errors derived as above may be used, but usually it is not certain that the straight line relationship will continue to hold. Thus for the data shown in *Figure 29* the percentage of sand may decrease more or less rapidly at depths of more than 48 in or, possibly, it may begin to increase again. It would therefore be a dangerous procedure to estimate the percentage of sand at a depth of 80 in. The same remarks are true concerning predictions in general. Thus, some form of relationship might be fitted to the population data in *Figure 24*, but unless it is reasonably certain that changing conditions would not alter the form of this relationship it would be of little use for predictive purposes.

6.7 *Correlation coefficients*—It is now easier to interpret the meaning and significance of the correlation coefficient

$$r=\frac{\Sigma(x-\bar{x})(y-\bar{y})}{\sqrt{[\{\Sigma(x-\bar{x})^2\}\{\Sigma(y-\bar{y})^2\}]}}$$

In the analysis of variance testing the linear regression of y on x, the sum of squares due to x was $[\Sigma(x-\bar{x})(y-\bar{y})]^2/\Sigma(x-\bar{x})^2$ out of a total sum of squares $\Sigma(y-\bar{y})^2$. The regression therefore accounted for a proportion

$$\frac{[\Sigma(x-\bar{x})(y-\bar{y})]^2}{\Sigma(x-\bar{x})^2\;\Sigma(y-\bar{y})^2}=r^2$$

of the total variation in y. It is apparent that r indicates the linear association between the two variables. Various conclusions may be drawn immediately:

1 Since r^2 gives the proportion of the total variability accounted for, it is less than unity, and $-1<r<1$. If r takes either of the values 1 or -1, all of the variability in y can be accounted for by changes in x and a perfect linear association exists between the two variables.

2 The significance of a value of r may be found using a variance ratio. Since a proportion $(1 - r^2)$ of the total variation is not accounted for and the residual mean square is $1/(n-2)$ times this, the variance ratio is

$$\frac{r^2}{(1 - r^2)/(n - 2)} = \frac{(n - 2)\, r^2}{1 - r^2}$$

This may be tested as a variance ratio with 1 and $n-2$ degrees of freedom. Alternatively the square root of this value

$$r\sqrt{\left(\frac{n-2}{1-r^2}\right)}$$

may be tested using the t table*.

For example, the correlation coefficient for the 65 pairs of observations shown in *Figure 23* is 0·487. The corresponding value for the variance ratio with 1 and 63 degrees of freedom is

$$\frac{63\,(0{\cdot}487)^2}{1 - (0{\cdot}487)^2} = 19{\cdot}6$$

This value is very highly significant and would arise by chance less than once in a hundred times.

3 Since the square of the correlation coefficient gives the proportion of the total variation that may be removed by a straight line relationship while the square of the correlation ratio gives the proportion removed by a general relationship, the correlation coefficient cannot exceed the correlation ratio. This does not mean that it is less sensitive than the correlation ratio in testing associations. On the contrary since the correlation coefficient employs only one degree of freedom it is usually more sensitive. However, since a more general relationship than a straight line might represent the data better, this might be tested using the analysis of variance.

For instance, the correlation coefficient for the data of *Figure 25* is 0·795 (compared with the correlation ratio of 0·829). The analysis of variance for a regression of the intelligence quotients on the degree of concentration is as follows:

	D.f.	*S.s.*	*M.s.*	*V.r.*
Variation ascribable to linear effect of concentration	1	5,385	5,385	53·1
Residual variation	31	3,145	101·5	
Total	32	8,530		

* Fisher and Yates have tabulated exact significance levels of r for each value of $n-2$. These tables may be used to determine directly the significance of any observed correlation coefficient.

There is no doubt about the significance of the regression. This analysis may now be combined with that carried out in section 6.3 in the following manner:

	D.f.	*S.s.*	*M.s.*
Variation ascribable to linear effect of concentration	1	5,385	5,385
Variation ascribable to non-linear effect of concentration	3	473	157·7
Between degrees of concentration	4	5,858	—
Residual	28	2,672	95·4
Total	32	8,530	

It is apparent that most of the variation ascribable to degree of concentration is removed by the straight line. The residual variation ascribable to possible non-linear effects of concentration is not significant so that no great improvement can be expected from further curve fitting. The failure of the non-linear term to reach significance does not, however, rule out the possibility that individual degrees of freedom, testing particular types of deviation from linearity, may reach significance.

Two points concerning the use of the correlation coefficient must be noted. First, the observations in each set are assumed to be independent of one another *i.e.* $x_1, x_2, \ldots x_n$ are assumed to be independent of one another and $y_1, y_2, \ldots y_n$ are assumed to be independent of one another. This means in effect that if the correlation coefficient is to be of general application and to apply to other data the observations must be independent. This is, of course, necessary for whatever purpose a set of observations may be used. If, for example, in observing the weights of children of a given age we restrict ourselves to a particular district or class then the weights so obtained will be applicable only to that district or class. It is, however, particularly necessary to note this when the association between two variables is being considered, since it is very easy to overlook the limitations of each set of observations. The correlation in *Figure* 29 is thus restricted to the range of observed depths of this particular soil profile. In *Figure* 22b the numbers of cows in milk at the June census in successive years are not independent measurements nor is the milk production in successive years since many of the same animals are observed in both years. This dependence of successive observations is reflected by the tendency of the points of the scatter diagram to fall in order. In consequence, the effect is as if some observations had been repeated and the apparent association is of limited interest and doubtful significance.

A second point of interest in testing the significance of the correlation coefficient is that when the number of pairs of observations is large it is possible to judge rapidly whether it is significantly different from zero using a standard error of $1/\sqrt{(n-1)}$. For example, for a correlation coefficient based on 65 pairs of observations this standard error is $1/\sqrt{64}=\pm 0{\cdot}125$. Thus, if $P=O$, the sample correlation coefficient r would exceed the value $0{\cdot}125\times 1{\cdot}96=0{\cdot}245$ by chance less than once in twenty times and the value $0{\cdot}125\times 3{\cdot}29=0{\cdot}411$ would be exceeded by chance less than once in a thousand times. The value 0·487 for the observations in *Figure 23* is hence very significant.

In general, since this is an approximate rule, it is most useful for gauging rapidly the significance of an observed correlation. Where there is any doubt as to the significance the exact test should be used.

SUMMARY OF PP 102 TO 118

Methods of measuring and testing the association between two sets of measurements have been given. A general test for association can be made using the analysis of variance or correlation ratio, but a more useful test is provided by the correlation coefficient

$$r=\frac{\Sigma(x-\bar{x})(y-\bar{y})}{\sqrt{[\{\Sigma(x-\bar{x})^2\}\{\Sigma(y-\bar{y})^2\}]}}$$

This tests the linear association between two series of observations. The significance of a correlation coefficient may be tested by employing $r^2(n-2)/(1-r^2)$ as a variance ratio with 1 and $n-2$ degrees of freedom.

It has been shown how straight lines may be fitted to estimate the dependence of one variable upon another variable. The equation estimating the dependence of one variable y upon another x is

$$y-\bar{y}=b(x-\bar{x})$$

where

$$b=\Sigma(x-\bar{x})(y-\bar{y})/\Sigma(x-\bar{x})^2$$

The significance of this may be tested using the analysis of variance.

EXAMPLES

52 The data of *Figure 24*a are given in the following table:

x	*Milk yield in first month*	20·2	20·4	22·5	25·7	25·8	25·9	27·7	28·1	28·3	29·4	29·6
y	*Milk yield in second month*	17·2	22·3	19·5	24·2	21·0	23·9	24·5	27·5	24·4	28·7	23·3
	Milk yield in first month	30·3	30·4	32·3	32·7	35·0	37·1	37·4	38·1	39·1		
	Milk yield in second month	26·4	30·3	29·4	28·7	33·2	31·7	32·1	38·1	32·9		

By grouping the data as indicated calculate the following analysis of variance to test the dependence of the mean daily milk yield in the second month of lactation upon that in the first month:

	D.f.	*S.s.*	*M.s.*	*V.r.*
Between groups	3	433·45	144·48	26·1
Within groups	16	88·66	5·54	
Total	19	522·11		

53 Show that for the data of the last example $r = 0{\cdot}918$ and that the mean milk yield in the second month of lactation may be estimated by the formula

$$26{\cdot}965 + 0{\cdot}8560\,(x - 29{\cdot}80)$$

where x is the mean milk yield in the first month of lactation.

Construct the analysis of variance to test the significance of this regression:

	D.f.	*S.s.*	*M.s.*	*V.r.*
Variation due to linear regression	1	439·85	439·85	96·2
Residual	18	82·26	4·57	
Total	19	522·11		

Here the linear regression is as effective as the crude grouping used in the last example and there is therefore no point in using a more complicated relationship.

Use the above formula and analysis to estimate the mean milk yield in the second month of lactation as $32{\cdot}27 \pm 0{\cdot}72$ lb when the mean daily milk yield in the first month is 36 lb.

54 Show that the standard error of the regression coefficient calculated in the last example is $\pm 0{\cdot}087$. Hence conclude that the regression coefficient does not differ significantly from unity and that the mean milk yield in the second month may be alternatively, but less efficiently, estimated by subtracting 2·84 lb from the mean yield of the first month.

55 The following figures give the heights in cm of male children on their fourth and fifth birthdays:

4th year	*5th year*	*4th year*	*5th year*	*4th year*	*5th year*
100·0	105·5	90·0	91·6	102·3	109·0
95·1	101·5	99·0	101·1	94·5	99·6
103·3	110·0	101·5	109·5	103·0	110·5
98·2	104·5	97·0	105·0	94·3	100·1
98·8	104·8	93·8	100·0	96·5	101·9
103·0	109·0	98·7	105·6	102·0	105·8
98·6	105·5	103·0	109·3	100·0	107·6
97·5	102·5	95·1	102·6	90·0	99·0
95·3	100·4	95·3	101·7	97·4	101·7
97·7	103·6	99·0	104·5	94·6	101·3
96·0	102·0	99·0	106·4	99·1	106·2
97·3	101·5	100·0	108·5	97·0	108·4
98·8	105·1	93·5	101·1	94·4	99·5
98·0	104·4	101·2	106·2	98·0	109·5
95·5	104·5	97·5	103·5	98·0	105·0
93·5	101·8	102·4	110·0	95·0	100·2
103·0	109·0	99·2	104·3	98·0	102·6
99·1	105·4	97·5	104·5	102·5	109·8
84·8	89·8	93·4	103·0	96·2	102·4
94·4	99·8	97·6	104·0	97·5	102·0

Plot a scatter diagram of these measurements and show that the correlation coefficient is 0·884

56 The following table gives the data plotted in *Figure 21a*:

Change in haemoglobin percentage	−14	−6	−5	−3	−3	−2	−2	−1
Change in weight	1·07	1·31	1·36	1·02	1·22	1·07	1·23	0·86
Change in haemoglobin percentage	2	2	4	6	6	6	7	7
Change in weight	1·06	0·62	1·08	0·66	0·92	1·39	0·71	0·74

Show that the correlation coefficient is −0·469 and hence conclude that as large a linear association between the variables would arise by chance more than once in twenty times but less than once in ten times.

57 For a series of 32 pairs of observations, the sums, and sums of squares and products were:

$$\Sigma x = 1{,}769 \qquad \Sigma y = 5{,}805$$

$$\Sigma (x-\bar{x})^2 = 22{,}683 \qquad \Sigma (x-\bar{x})(y-\bar{y}) = 37{,}937 \qquad \Sigma (y-\bar{y})^2 = 119{,}912$$

Show that a further observation $x=100$, $y=300$ might easily arise by chance.

EXTENDED DEVELOPMENT

6A.8 *Theory of minimal variance*—In section 6.5 certain results for the best estimates of the regression coefficient were quoted without proof. The derivation of these results will now be considered.

We have to choose the unknown values in the regression equation so that the residual variation is a minimum. In terms of the notation of section 6.5, it is required to minimize $\Sigma[(y-y_0)-\beta(x-x_0)]^2$. In particular, choose $x_0=\bar{x}$ and minimize $\Sigma[(y-y_0)-\beta(x-\bar{x})]^2$.

This may be done by differential calculus and the following equations are obtained for y_0 and β:

$$\Sigma[(y-y_0)-\beta(x-\bar{x})]=0 \qquad \text{i.e. } \Sigma y - n y_0 = 0 \qquad \text{i.e. } y_0=\bar{y}$$

$$\Sigma(x-\bar{x})[(y-y_0)-\beta(x-\bar{x})]=0 \qquad \text{i.e. } \Sigma(x-\bar{x})(y-y_0)=\beta\Sigma(x-\bar{x})^2$$

$$\text{i.e. } \beta=\frac{\Sigma(x-\bar{x})(y-\bar{y})}{\Sigma(x-\bar{x})^2}$$

These quantities are only estimates of the true values so that we write

$$b=\frac{\Sigma(x-\bar{x})(y-\bar{y})}{\Sigma(x-\bar{x})^2}$$

Now $y-y_0=\beta(x-\bar{x})+e$, so that

$$b=\frac{\Sigma(x-\bar{x})[y_0+\beta(x-\bar{x})+e-\bar{y}]}{\Sigma(x-\bar{x})^2}$$

$$=\frac{\Sigma(x-\bar{x})(y_0-\bar{y})}{\Sigma(x-\bar{x})^2}+\frac{\Sigma\beta(x-\bar{x})^2}{\Sigma(x-\bar{x})^2}+\frac{\Sigma(x-\bar{x})e}{\Sigma(x-\bar{x})^2}$$

Since $\Sigma(x-\bar{x})(y_0-\bar{y})=(y_0-\bar{y})\Sigma(x-\bar{x})=0$, the first term disappears and we get

$$b=\beta+\frac{\Sigma(x-\bar{x})e}{\Sigma(x-\bar{x})^2}$$

or

$$b-\beta=\frac{\Sigma(x-\bar{x})e}{\Sigma(x-\bar{x})^2}$$

This gives the amount by which the estimated coefficient b will differ from the true coefficient β. It will, on average, be zero. The variance of b may now be found by considering the variance of the right hand side of this equation. Writing this out fully it is

$$\frac{(x_1-\bar{x})}{\Sigma(x-\bar{x})^2}e_1+\frac{(x_2-\bar{x})}{\Sigma(x-\bar{x})^2}e_2+\ldots\ldots\ldots+\frac{(x_n-\bar{x})}{\Sigma(x-\bar{x})^2}e_n$$

If σ^2 is the variance of e, the variance of the first term is

$$\frac{(x_1-\bar{x})^2}{[\Sigma(x-\bar{x})^2]^2}\sigma^2$$

of the second term is

$$\frac{(x_2-\bar{x})^2}{[\Sigma(x-\bar{x})^2]^2}\sigma^2$$

and so on. Thus the variance of the whole expression is

$$\frac{(x_1-\bar{x})^2}{[\Sigma(x-\bar{x})^2]^2}\sigma^2+\frac{(x_2-\bar{x})^2}{[\Sigma(x-\bar{x})^2]^2}\sigma^2+\ldots\ldots+\frac{(x_n-\bar{x})^2}{[\Sigma(x-\bar{x})^2]^2}\sigma^2$$

$$=\frac{\sigma^2}{[\Sigma(x-\bar{x})^2]^2}[(x_1-\bar{x})^2+(x_2-\bar{x})^2+\ldots\ldots+(x_n-\bar{x})^2]$$

$$=\frac{\sigma^2}{[\Sigma(x-\bar{x})^2]^2}\times\Sigma(x-\bar{x})^2$$

$$=\frac{\sigma^2}{\Sigma(x-\bar{x})^2}$$

This proves the third part of the formula quoted in section 6.5.

This method of minimizing the variance is also called the method of least squares. Some general results that may be obtained by this method will now be considered.

If there are several sets of observations, $y_1, y_2, \ldots y_n$; $x_1, x_2, \ldots x_n$; $t_1, t_2, \ldots t_n$; and it is required to estimate the dependence of the first set upon the others, a general linear relationship might be fitted of the form

$$y=y_0+\beta_1(x-\bar{x})+\beta_2(t-\bar{t})+\ldots$$

This is called a multiple regression. Here a unit increase in x causes an increase of β_1 in y, a unit increase in t causes an increase of β_2 in y *etc.*

For example, it may be possible to estimate weight from height but this estimate may be further improved by taking account of chest or waist circumference or, perhaps, leg length. A linear equation might then be used to estimate weight from other measurements. First, however, it is necessary to estimate the unknown coefficients of the equation. This may be done by the method of least squares.

This method gives $\bar{y}$ as the estimate of y_0 and the regression coefficients may be estimated from the equations:

$$\Sigma (x-\bar{x}) [y-\bar{y}-b_1 (x-\bar{x})-b_2 (t-\bar{t})- \dots]=0$$

$$\Sigma (t-t) [y-\bar{y}-b_1 (x-\bar{x})-b_2 (t-\bar{t})- \dots]=0$$

etc

i.e. from the equations:

$$b_1 \Sigma (x-\bar{x})^2 + b_2 \Sigma (x-\bar{x})(t-\bar{t}) + \dots = \Sigma (x-\bar{x})(y-\bar{y})$$

$$b_1 \Sigma (x-\bar{x})(t-\bar{t}) + b_2 \Sigma (t-\bar{t})^2 + \dots = \Sigma (t-\bar{t})(y-\bar{y})$$

etc

These equations have to be solved to estimate the regression coefficients.

Here the sum of squares that might be attributed to x, t, ... in the analysis of variance for y may be shown by lengthy algebra to be

$$\Sigma (y-\bar{y}) [b_1 (x-\bar{x}) + b_2 (t-\bar{t}) + \dots]$$

i.e.

$$b_1 \Sigma (x-\bar{x})(y-\bar{y}) + b_2 \Sigma (t-\bar{t})(y-\bar{y}) + \dots$$

with as many degrees of freedom as there are independent variables. This provides a joint test of the association between the dependent variate y and the independent variates $x, t \dots$, but it must be noted that the individual terms in the above expression do not test anything; only the total expression may be used in the analysis of variance.

A second point that should be noted is that the regression coefficients of y on x, t, ... may be altered by the inclusion of another independent variable. This is fairly obvious if an example is taken. Consider an equation to estimate height from weight and chest circumference. If a further variable is now introduced, say, leg length, then the entire structure of the equation may be altered. Since leg length is known, weight and chest circumference should now be used, in effect, to estimate trunk length. To take a second example: consider an equation to estimate the heights of 5-year old boys from their heights at three years. The inclusion of height at four years as an extra variable would obviously cause the height at three years to be of lesser importance. So the inclusion of any extra variable may completely change the form of a regression equation.

A last point to be noted is that the sum of squares in the analysis of variance for y will not usually be the sums of the individual contributions from x, t, ... *i.e.*

$$\frac{[\Sigma (x-\bar{x})(y-\bar{y})]^2}{\Sigma (x-\bar{x})^2} + \frac{[\Sigma (t-\bar{t})(y-\bar{y})]^2}{\Sigma (t-\bar{t})^2} + \ldots$$

The reason for this is again best illustrated by consideration of particular examples. Height at five years may be predicted using either height at four years or height at four years and one day. Both these predictions might be very reasonable, but by using both height at four years and height at four years and one day we shall not improve either much. There is a close correlation between height at four years and height at four years and one day and, in consequence, their joint use gives little or no improvement in the prediction. Sometimes, however, the reverse position will occur: a joint regression may account for more than the sum of the individual regressions. Here the joint regression may serve to eliminate some factor influencing both independent variables. For instance, suppose it is required to predict ability in Arithmetic. There might be two tests which measure, first, ability in English and, secondly, ability in English and Arithmetic. The first is of no use and the second is of limited use. However, these might be employed jointly to determine ability in Arithmetic fairly accurately. The two together can be used to eliminate the undesirable factor.

Thus, in general, it is not possible to predict what will happen in a multiple regression from a series of single regressions. We know that the predictive value of the multiple regression cannot be worse than the best of the single regressions, but it may be very much better than the individual regressions may seem to indicate.

If the correlation between the independent variables is zero,

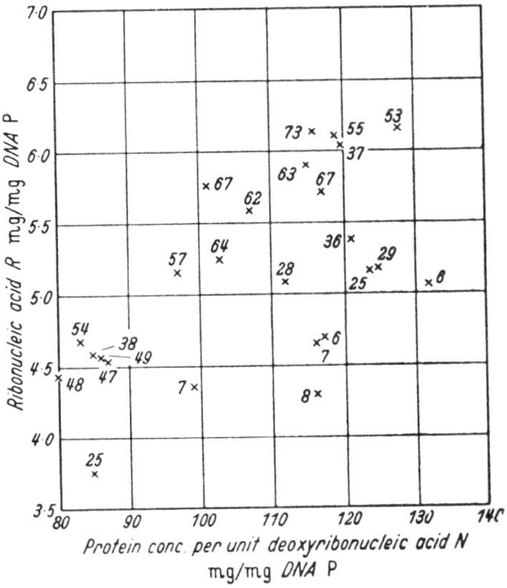

Figure 30. Plot of R against N

then the sums of squares for the individual regressions on each variable can be added to give the sum of squares for overall regression. The individual components here each contribute one degree of freedom to the total and may be used to test the significance of each variate.

6A.9 *Example of multiple regression*—The following example [data of CAMPBELL, R. M., and KOSTERLITZ, H. W. *J. Endocrinology* 6 (1949) 171] demonstrates the application of the above theory.

Estimates were made of the ribonucleic acid R and protein N concentrations expressed in terms of the concentration of deoxyribonucleic acid in the livers of 25 pregnant rats. Further observations were also taken on the weight W gm of uterus for each animal. *Figure 30* shows a plot of R against N, the values of W being indicated against each point. It is seen that the dependence of R on N might be represented by a straight line. However, it is also very obvious that the animals with a low uterine weight have a low value of R. Consider therefore a joint linear regression of R upon N and W.

The first step in the analysis is to calculate the sums, and sums of squares and products:

$$\Sigma N = 2{,}691 \qquad \Sigma W = 1{,}011 \qquad \Sigma R = 128{\cdot}06$$

$$\Sigma (N-\bar{N})^2 = 6{,}040 \qquad \Sigma (N-\bar{N})(W-\bar{W}) = -1{,}857 \qquad \Sigma (N-\bar{N})(R-\bar{R}) = 150{\cdot}2$$

$$\Sigma (W-\bar{W})^2 = 11{,}462 \qquad \Sigma (W-\bar{W})(R-\bar{R}) = 206{\cdot}0$$

$$\Sigma (R-\bar{R})^2 = 10{\cdot}72$$

The regression coefficients b_N and b_W may now be estimated from the equations:

$$6{,}040\, b_N - 1{,}857\, b_W = 150{\cdot}2$$

$$-1{,}857\, b_N + 11{,}462\, b_W = 206{\cdot}0$$

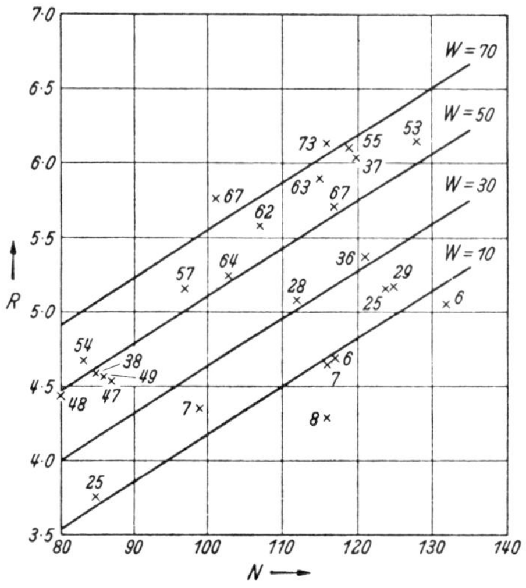

Figure 31. Regression lines fitted to data of Figure 30

giving

$$b_N = 0{\cdot}031987$$

$$b_W = 0{\cdot}023155$$

The means are $\bar{R} = 5{\cdot}122$, $\bar{N} = 107{\cdot}64$, $\bar{W} = 40{\cdot}44$ so that the regression equation is

$$R = 5{\cdot}122 + 0{\cdot}031987\,(N - 107{\cdot}64) + 0{\cdot}023155\,(W - 40{\cdot}44)$$

This cannot be represented by a single line in *Figure 30* since the values of W vary. One possible form of representation is shown in *Figure 31*. Here, a series of regression lines is shown for different values of W.

The sum of squares due to the regression is

$$0.031987 \times 150{\cdot}2 + 0{\cdot}023155 \times 206{\cdot}0 = 9{\cdot}57$$

so the analysis of variance for R is as follows:

	D.f.	*S.s.*	*M.s.*	*V.r.*
Variation due to regression	2	9·57	4·785	91·5
Residual	22	1·15	0·0523	
Total	24	10·72		

The regression is highly significant. It should be noted that the individual regressions of R on N and R on W are both significant, accounting for $(150{\cdot}2)^2/6{,}040=3{\cdot}74$ and $(206{\cdot}0)^2/11{,}462=3{\cdot}70$ of the total sum of squares respectively. These two values total to 7·44 which is less than the value 9·57 for the joint regression. Evidently when considered jointly the two variables account for a much greater proportion of the total variation in R. This difference occurs since N and W are negatively correlated ($r=-0{\cdot}22$) and the inclusion of both greatly reduces the unaccountable variability in R.

The proportion of the total variability in any measurement that can be accounted for by a multiple regression can be found using the analysis of variance. Here again the square root of this proportion may be employed; this quantity is called the multiple correlation coefficient and is usually denoted by R.

In the last example N and W account for a proportion, $9{\cdot}57/10{\cdot}72=0{\cdot}893$, of the total variability so that the multiple correlation coefficient will be $\sqrt{(0{\cdot}893)}=0{\cdot}945$.

6A.10 *Significance of particular variates*—A complete account of the methods and tests of multiple regression is beyond the scope of this book but there is one particular test which is often required. It is frequently necessary to test whether the inclusion of an independent variable in a multiple regression is worth while. This is not the same as testing whether the independent variable is correlated with the dependent variable for it may be correlated but still be unable to contribute anything extra to the independent variables already employed. For example, when height at four years has already been used in estimating height at five years, it is doubtful if the use of height at four years and one day would lead to an improved prediction.

The analysis of variance may again be used in testing the significance of the improvement in the prediction. The increase in the regression sum of squares due to the inclusion of the extra variate may be tested by the variance-ratio test. For the example of the last section the analysis of variance for R when a regression is carried out on N alone is:

	D.f.	*S.s.*	*M.s.*	*V.r.*
Variation due to N	1	3·74	3·74	12·3
Residual variation	23	6·98	0·303	
Total	24	10·72		

The joint regression of R on N and W accounts for a sum of squares of 9·57, so that the analysis of variance may be written:

	D.f.	S.s.	M.s.	V.r.
Variation due to N	1	3·74	3·74	71·5
Extra variation accounted for by W	1	5·83	5·83	111·5
Variation due to N and W	2	9·57	—	—
Residual variation	22	1·15	0·0523	
Total	24	10·72		

The extra variation that may be accounted for by W is highly significant.

When the effects of one or more variables have been eliminated, it is convenient to consider the proportion of the residual variation that may be accounted for by including a further variable. Here the inclusion of W accounts for 5·83 of the residual variation 6·98 so that the proportion is $5{\cdot}83/6{\cdot}98=0{\cdot}835$. The square root of this value $\sqrt{(0{\cdot}835)}=0{\cdot}914$ is called the partial correlation coefficient of R and W, when N has been eliminated.

In the above example there is little doubt about the significance of both W and N, since the joint regression accounts for a high proportion of the total variation in R, but in experimentation there may often be some doubt as to the usefulness of some measurements. The following example illustrates this:

Measurements were taken of the weights W, standing heights H, sitting heights S, and chest circumferences C of twenty 3-year old children. These gave the following values:

$\Sigma H=18{,}263$	$\Sigma S=10{,}863$	$\Sigma C=10{,}180$	$\Sigma W=567{\cdot}4$
$\Sigma(H-\bar{H})^2$ $=46{,}997$	$\Sigma(H-\bar{H})(S-\bar{S})$ $=20{,}239$	$\Sigma(H-\bar{H})(C-\bar{C})$ $=4{,}976$	$\Sigma(H-\bar{H})(W-\bar{W})$ $=1{,}397{\cdot}6$
	$\Sigma(S-\bar{S})^2$ $=12{,}725$	$\Sigma(S-\bar{S})(C-\bar{C})$ $=4{,}300$	$\Sigma(S-\bar{S})(W-\bar{W})$ $=629{\cdot}7$
		$\Sigma(C-\bar{C})^2$ $=8{,}282$	$\Sigma(C-\bar{C})(W-\bar{W})$ $=406{\cdot}8$
			$\Sigma(W-\bar{W})^2$ $=95{\cdot}82$

Suppose it is necessary to predict W using H, S and C. The individual regressions of W on H, S and C in turn account for 41·56, 31·16 and 19·98 of the total sum of squares and each variable is thus worth consideration. The regression coefficients in the overall regression on the three variables are calculated from:

$$46{,}997\,b_H+20{,}239\,b_S+4{,}976\,b_C=1{,}397{\cdot}6$$

$$20{,}239\,b_H+12{,}725\,b_S+4{,}300\,b_C=629{\cdot}7$$

$$4{,}976\,b_H+4{,}300\,b_S+8{,}282\,b_C=406{\cdot}8$$

giving

$$b_H=0{\cdot}031401 \qquad b_S=-0{\cdot}012954 \qquad b_C=0{\cdot}036978$$

The sum of squares due to the regression is thus

$$0{\cdot}031401 \times 1{,}397{\cdot}6 - 0{\cdot}012954 \times 629{\cdot}7 + 0{\cdot}036978 \times 406{\cdot}8 = 50{\cdot}77$$

and the analysis of variance is:

	D.f.	*S.s.*	*M.s.*	*V.r.*
Variation attributable to regression	3	50·77	16·92	6·01
Residual variation	16	45·05	2·816	
Total	19	95·82		

This is obviously significant, but the sum of squares is not very much larger than that due to H alone so the analysis may be written:

	D.f.	*S.s.*	*M.s.*	*V.r.*
Variation attributable to H alone	1	41·56	41·56	14·8
Extra variation attributable to S and C	2	9·21	4·605	1·64
Variation attributable to regression	3	50·77	—	—
Residual variation	16	45·05	2·816	
Total	19	95·82		

The improvement due to the inclusion of S and C is not significant. It is also apparent that neither of these could contribute significantly to the regression since, even if the whole 9·21 could be accounted for by either variable, it would still not be significant.

6A.11 *Partial correlation coefficients*—The partial correlation coefficient was introduced in the last section as indicating the proportion of the variation in one variable that could be accounted for by changes in another when the effects of other variables had been eliminated. As such it is similar to the correlation coefficient and may be tested in a similar manner.

If r denotes the partial correlation coefficient, and m variables have been eliminated, then $r^2(n-m-2)/(1-r^2)$ can be tested as a variance ratio with 1 and $n-m-2$ degrees of freedom. This would be of little interest if it were not possible to calculate a partial correlation coefficient without using the analysis of variance. It is, however, possible to calculate a partial correlation coefficient using the correlation coefficients of each pair of variables. This is often very useful in determining which variables are worth consideration.

If r_{xy} denotes the correlation coefficient between variables x and y, r_{xz} the correlation coefficient between variables x and z and so on, the partial correlation coefficient between x and y when the effect of z has been eliminated is given by

$$r_{xy\,.\,z} = \frac{r_{xy} - r_{xz}\, r_{yz}}{\sqrt{[(1-r_{xz}^2)(1-r_{yz}^2)]}}$$

The same formula may be used for partial correlations where several variables are eliminated. For example, the partial correlation coefficient between x and y when the effect of z and t has been eliminated may be calculated using

$$r_{xy.\,zt}=\frac{r_{xy.\,t}-r_{xz.\,t}\,r_{yz.\,t}}{\sqrt{[(1-r_{xz.\,t}^{2})(1-r_{yz.\,t}^{2})]}}$$

where each partial correlation in this formula may be calculated using formulae of the above type. The data from the example of the last section may be used to demonstrate these formulae.

From the sums of squares and products the following correlation coefficients may be calculated:

$r_{HS}=0{\cdot}8276$ $r_{HC}=0{\cdot}2522$ $r_{SC}=0{\cdot}4189$ $r_{HW}=0{\cdot}6586$ $r_{SW}=0{\cdot}5703$ $r_{CW}=0{\cdot}4567$

These coefficients show that weight is significantly correlated with standing height, sitting height and chest circumference. The correlation with standing height is greatest so this variable might be eliminated to give:

$$r_{SW.\,H}=\frac{0{\cdot}5703-0{\cdot}8276\times0{\cdot}6586}{\sqrt{[(1-0{\cdot}8276^{2})\,(1-0{\cdot}6586^{2})]}}=0{\cdot}0597$$

$$r_{CW.\,H}=\frac{0{\cdot}4567-0{\cdot}2522\times0{\cdot}6586}{\sqrt{[(1-0{\cdot}2522^{2})\,(1-0{\cdot}6586^{2})]}}=0{\cdot}3991$$

$$r_{SC.\,H}=\frac{0{\cdot}4189-0{\cdot}8276\times0{\cdot}2522}{\sqrt{[(1-0{\cdot}8276^{2})\,(1-0{\cdot}2522^{2})]}}=0{\cdot}3869$$

The correlation between sitting height and weight has been greatly reduced by elimination of the effect of standing height and is obviously not significant. The partial correlation coefficient $r_{CW.\,H}$ may be tested using the variance ratio

$$\frac{17\,(0{\cdot}3991)^{2}}{1-0{\cdot}3991^{2}}=3{\cdot}22$$

with 1 and 17 degrees of freedom. This does not reach significance so that there is no point in further analysis. If, however, it had reached significance, C might then have been eliminated and $r_{SW\,.\,HC}$ calculated and tested thus:

$$r_{SW.\,HC}=\frac{0{\cdot}0597-0{\cdot}3991\times0{\cdot}3869}{\sqrt{[(1-0{\cdot}3991^{2})\,(1-0{\cdot}3869^{2})]}}=-0{\cdot}1120$$

Successive steps of this type allow us to pick the variables of greatest significance, but we cannot decide which variables might be neglected until the analysis is complete since insignificant variables at one stage in the analysis may become significant at the next elimination.

6A.12 *Curvilinear regression*—So far the fitting of straight lines only has been considered because a linear relationship will usually represent fairly adequately any association between two variables. However, a straight line may not always suffice, and more complicated forms of representation may have to be used. Examples have already been given in *Figures 24* and *26* of non-linear or curvilinear associations.

If an association is not linear then it is most commonly represented by a polynomial, but still more complicated forms of representation involving logarithms or exponentials may be required to specify the form of dependence of one variable upon others.

If the dependence of one variable y on another x is curved then the simplest polynomial representation involves a term in x^2 and is of the form

$$y=a+bx+cx^{2}$$

Examples of shapes of such curves are shown in *Figure 32*a.

a *Curves of second degree*

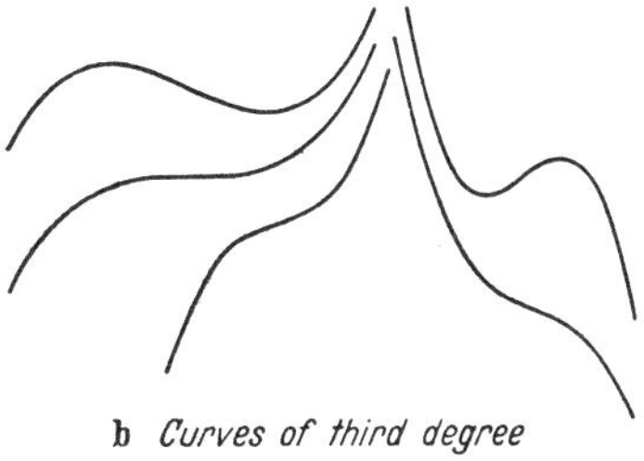

b *Curves of third degree*

Figure 32

Although each of these has a maximum or minimum, it is not necessary for the original data to have a maximum or minimum for this curve to be a valid representation over part of the range.

If a further term in x^3 is included giving a relationship of the form

$$y=a+bx+cx^2+dx^3$$

then the curve takes shapes similar to those shown in *Figure 32*b. Inclusion of more terms leads to correspondingly more complex and flexible relationships.

When the form of the curve which it is desired to fit has been settled, the coefficients in its equation have to be estimated. Fortunately this estimation does not involve any new principles. Coefficients may be estimated in exactly the same manner as for a multiple regression if x, x^2, x^3, ... are considered as each being different variates. The regression equation might be written

$$y=y_0+\beta_1(x-\bar{x})+\beta_2(x^2-\overline{x^2})+\beta_3(x^3-\overline{x^3})+\ldots$$

where $\overline{x^2}$ is the mean value of x^2, $\overline{x^3}$ is the mean value of x^3, and so on. The unknown coefficients may then be estimated in the usual manner from the equations:

$$y_0=\bar{y}$$

$$b_1\Sigma(x-\bar{x})^2+b_2\Sigma(x-\bar{x})(x^2-\overline{x^2})+b_3\Sigma(x-\bar{x})(x^3-\overline{x^3})+\ldots=\Sigma(x-\bar{x})(y-\bar{y})$$

$$b_1\Sigma(x-\bar{x})(x^2-\overline{x^2})+b_2\Sigma(x^2-\overline{x^2})^2+b_3\Sigma(x^2-\overline{x^2})(x^3-\overline{x^3})+\ldots=\Sigma(x^2-\overline{x^2})(y-\bar{y})$$

$$b_1\Sigma(x-\bar{x})(x^3-\overline{x^3})+b_2\Sigma(x^2-\overline{x^2})(x^3-\overline{x^3})+b_3\Sigma(x^3-\overline{x^3})^2+\ldots=\Sigma(x^3-\overline{x^3})(y-\bar{y})$$

Here each sum of products may be calculated in the usual manner *e.g.*

$$\Sigma(x^2-\overline{x^2})(x^3-\overline{x^3})=\Sigma x^2 . x^3-\frac{(\Sigma x^2)(\Sigma x^3)}{n}$$

$$=\Sigma x^5-\frac{(\Sigma x^2)(\Sigma x^3)}{n}$$

$$\Sigma(x^3-\overline{x^3})(y-\bar{y})=\Sigma x^3 y-\frac{(\Sigma x^3)(\Sigma y)}{n}$$

but the sums of powers of x have now to be calculated. The calculation of these may be shortened by subtracting a value roughly equal to the mean from each observation of x before carrying out the calculation. The same approach may usually be employed when more complicated terms such as $\log x$, e^x or $\sin x$ occur in the regression equation.

If, however, a polynomial is being fitted, it is of interest to test whether successive terms in x^2, x^3 and so on contribute to the regression. This may be done by a suitable arrangement of the calculation. The following example will demonstrate the method:

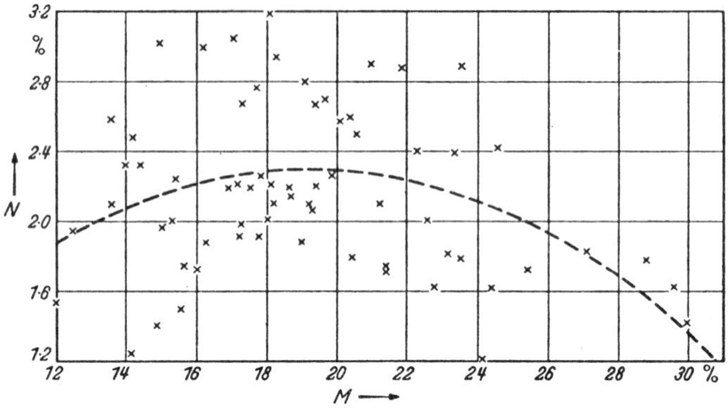

Figure 33. Plot of percentage nitrogen N in dry matter against percentage dry matter M in silage

A series of 67 observations (unpublished data of A. J. BARNETT) was taken of the nitrogen content of the dry matter in silage N and of the percentage dry matter M. These are plotted in *Figure 33*. There is apparently a maximum indicating that a straight line fit would probably not suffice and a term in M^2 may be necessary. In addition, this hump does not seem to be symmetrical indicating that a term in M^3 might lead to a significant improvement in fit. Thus a cubic equation might be fitted to the data.

The first step is to calculate the sums of the powers of M, but this may be shortened by subtracting, say, 20 from each observation and using $m = M - 20$ instead of M. This gives the values:

$\overline{m} = -0{\cdot}83 \qquad \overline{m^2} = 16{\cdot}64 \qquad \overline{m^3} = 2{\cdot}57 \qquad \overline{N} = 2{\cdot}166$

$\Sigma(m-\overline{m})^2 = 1{,}073 \qquad \Sigma(m-\overline{m})(m^2-\overline{m^2}) = 1{,}095 \qquad \Sigma(m-\overline{m})(m^3-\overline{m^3}) = 49{,}715 \qquad \Sigma(m-\overline{m})(N-\overline{N}) = -21{\cdot}619$

$\Sigma(m^2-\overline{m^2})^2 = 30{,}956 \qquad \Sigma(m^2-\overline{m^2})(m^3-\overline{m^3}) = 135{,}618 \qquad \Sigma(m^2-\overline{m^2})(N-\overline{N}) = -263{\cdot}650$

$\Sigma(m^3-\overline{m^3})^2 = 3{,}243{,}023 \qquad \Sigma(m^3-\overline{m^3})(N-\overline{N}) = -1{,}277{\cdot}956$

$\Sigma(N-\overline{N})^2 = 15{\cdot}000$

The equations to estimate the regression coefficients are thus:

$$1{,}073\, b_1 + 1{,}095\, b_2 + 49{,}715\, b_3 = -21{\cdot}619$$
$$1{,}095\, b_1 + 30{,}956\, b_2 + 135{,}618\, b_3 = -263{\cdot}650$$
$$49{,}715\, b_1 + 135{,}618\, b_2 + 3{,}243{,}023\, b_3 = -1{,}277{\cdot}956$$

The sum of squares due to the linear component of m is $(-21{\cdot}619)^2/1{,}073 = 0{\cdot}436$. This value is small, but the linear term in m should be included if higher powers are to be used.

If b_1 is now eliminated from the second and third equations by subtracting 1,095/1,073 and 49,715/1,073 times the first equation from these respectively, then:

$$29{,}839\ b_2 + 84{,}884\ b_3 = -241{\cdot}588$$

$$84{,}884\ b_2 + 939{,}592\ b_3 = -276{\cdot}289$$

The increase in the regression sum of squares due to the inclusion of a quadratic term when the linear term has already been employed is then $(-241{\cdot}588)^2/29{,}839 = 1{\cdot}956$. Next b_2 is eliminated from the third equation by subtracting 84,884/29,839 times the second equation to give

$$698{,}120\ b_3 = 410{\cdot}964$$

The increase in the regression sum of squares due to the inclusion of a cubic term when the linear and quadratic terms have been employed is $(410{\cdot}964)^2/698{,}120 = 0{\cdot}242$. The analysis of variance is:

	D.f.	*S.s.*	*M.s.*	*V.r.*
Variation due to linear term	1	0·436	0·436	2·22
Extra variation due to quadratic term	1	1·956	1·956	9·96
Extra variation due to cubic term	1	0·242	0·242	1·23
Total variation due to regression	3	2·634	——	——
Residual variation	63	12·366	0·1963	
Total	66	15·000		

The total variation due to the regression may be checked by completing the solution for b_1, b_2 and b_3 to give

$$b_1 = -0{\cdot}0374516 \qquad b_2 = -0{\cdot}0097710 \qquad b_3 = 0{\cdot}0005887$$

and the sum of squares is

$$-0{\cdot}0374516 \times (-21{\cdot}619) - 0{\cdot}0097710 \times (-263{\cdot}650) + 0{\cdot}0005887 \times (-1277{\cdot}956) = 2{\cdot}634$$

Here, however, the improvement due to the inclusion of a cubic term is not significant so that b_3 should be dropped from the regression equations. This gives:

$$1{,}073\ b_1 + 1{,}095\ b_2 = -21{\cdot}619$$

$$1{,}095\ b_1 + 30{,}956\ b_2 = -263{\cdot}650$$

$$29{,}839\ b_2 = -241{\cdot}588$$

$$b_2 = -0{\cdot}0080964$$

$$b_1 = -0{\cdot}0118858$$

The corresponding sum of squares due to the regression is now

$$-0{\cdot}0118858 \times (-21{\cdot}619) - 0{\cdot}0080964 \times (-263{\cdot}650) = 2{\cdot}392$$

The analysis of variance may next be completed thus:

	D.f.	*S.s.*	*M.s.*	*V.r.*
Variation due to linear term	1	0·436	0·436	——
Extra variation due to quadratic term	1	1·956	1·956	——
Total variation due to regression	2	2·392	1·196	6·07
Residual variation	64	12·608	0·1970	
Total	66	15·000		

This quadratic regression is significant at the 1 per cent level. We may now calculate the regression equation by setting $m=M-20$. This gives the estimated value N as

$$2{\cdot}166-0{\cdot}0118858(m+0{\cdot}83)-0{\cdot}0080964(m^2-16{\cdot}64)=-0{\cdot}71+0{\cdot}31197M-0{\cdot}0080964M^2$$

The dotted line in *Figure 33* shows this fitted curve. Here, more than for the straight line, it is dangerous to extrapolate beyond the range of the observed values. It cannot be concluded that for a percentage dry matter of 36 the mean percentage nitrogen is 0·03. It may only be concluded that within the range observed there is a non-linear association which may be represented by the above equation. From these observed values the peak could not be said to be symmetrical. The inclusion of a cubic term does not provide a better representation, but further observations might easily make the cubic term significant.

The above technique of testing each extra term as it is introduced should be noted. If the calculation is carefully carried out the regression coefficients need not be calculated nor the sum of squares due to the regression checked until the order of the fitted polynomial has been decided. Thus the testing of successive terms may proceed step by step.

This step by step testing may be simplified still further if the values taken by the independent variate are equally spaced. The sums of squares and products which are the coefficients of the regression equations may be tabulated and the whole process of fitting greatly shortened. This gives rise to the method known as the fitting of orthogonal polynomials. A tabulation of these polynomials with an explanation of their use is given in FISHER, R. A., and YATES, F. *Statistical Tables for Biological, Agricultural and Medical Research* London, 1947.

SUMMARY OF PP 120 TO 132

The theory of fitting straight lines and curves has been given. It has been shown how one variable can depend on several others and methods of testing the significance of the dependence have been demonstrated. In particular, applications of the partial correlation coefficient

$$r_{xy\,.\,z}=\frac{r_{xy}-r_{xz}\,r_{yz}}{\sqrt{[(1-r_{xz}^2)(1-r_{yz}^2)]}}$$

have been considered and the appropriate tests of significance given.

EXAMPLES

58 Show that for a fitted straight line the sum of squares of deviations of observations from the straight line may be written in the forms shown:

$$\Sigma e^2=\Sigma[y-\bar{y}-b\,(x-\bar{x})]^2$$

$$=\Sigma(y-\bar{y})^2-\frac{[\Sigma\,(x-\bar{x})\,(y-\bar{y})]^2}{\Sigma\,(x-\bar{x})^2}$$

This provides another proof that the sum of squares accounted for by the straight line is $[\Sigma\,(x-\bar{x})\,(y-\bar{y})]^2/\Sigma\,(x-\bar{x})^2$.

59 The following table gives the unemployment in thousands in London and Wales during the twelve months of 1946:

Month M	1	2	3	4	5	6	7	8	9	10	11	12
London L	24·5	27·4	30·8	31·6	34·8	36·0	34·7	36·1	36·3	39·4	40·0	38·4
Wales W	68·2	69·8	69·6	68·8	66·9	66·3	61·1	60·1	57·7	56·4	54·0	53·2

Show that the following correlations exist:

$$r_{LW} = -0{\cdot}8237 \qquad r_{LM} = 0{\cdot}9346 \qquad r_{WM} = -0{\cdot}9625 \qquad r_{LW.M} = 0{\cdot}7872$$

These correlations show that proportions $(0{\cdot}9346)^2 = 0{\cdot}87$ and $(-0{\cdot}9625)^2 = 0{\cdot}93$ of the total variation in L and W can be accounted for by a linear trend with time. In consequence, L and W are negatively correlated. However, when the trend is removed from each variable the residuals are positively correlated.

This example demonstrates how it is possible to remove a trend from each of two variables and to test the correlation between the residual portions.

60 The following table gives the increase in the cross sectional area of a European larch over a series of years [data of LIANG, S. C. *Forestry* 22 (1949) 222]:

Year	1	2	3	4	5	6	7	8
Increase in area cm^2	2·5	4·9	5·6	8·4	7·6	6·9	10·8	9·4
Year	9	10	11	12	13	14	15	16
Increase in area cm^2	11·1	11·1	9·2	7·8	12·0	12·9	12·3	10·5

Fit and test a quadratic curve giving the annual rate of increase.

61 The following table gives the diameters x and depressions y of walled lunar craters:

Diameter x km	*Depression y m*	*Diameter x km*	*Depression y m*	*Diameter x km*	*Depression y m*
13	1,700	35	550	80	400
14	1,400	46	400	82	400
14	1,050	48	750	85	1,050
20	1,600	50	350	95	800
20	800	69	1,100	100	1,050
29	550	72	700	115	2,600
30	1,500	72	1,450	230	3,250
33	400				

Show that the dependence of y on x might be represented by a quadratic equation of the form

$$y = 1{,}185 - 10{\cdot}21x + 0{\cdot}0872x^2$$

62 A series of observations was made on 221 men noting their heights h, weights w, and metabolic rates m. The sums of squares and products were:

$$\Sigma(h-\bar{h})^2 = 10{,}926 \qquad \Sigma(h-\bar{h})(w-\bar{w}) = 6{,}953 \qquad \Sigma(h-\bar{h})(m-\bar{m}) = 139{,}456$$
$$\Sigma(w-\bar{w})^2 = 26{,}702 \qquad \Sigma(w-\bar{w})(m-\bar{m}) = 377{,}323$$
$$\Sigma(m-\bar{m})^2 = 6{,}289{,}060$$

Test the linear dependence of m on h and w and the significance of each variable.

7

CONCOMITANT OBSERVATIONS

7.1 *Comparison of regression coefficients*—Often when fitting several regression lines it is necessary to test whether their slopes are significantly different. For example it may be necessary to test whether the weight increase for each unit increase in height is the same for males and females. The problem, here, is essentially that of comparing regression coefficients.

Since from the last chapter the variance of a linear regression coefficient is $\sigma^2/\Sigma(x-\bar{x})^2$ regressions may be compared in exactly the same manner as means. Thus, if there are two regressions of y on x and y' on x' the two regression coefficients are estimated from

$$b=\frac{\Sigma(x-\bar{x})(y-\bar{y})}{\Sigma(x-\bar{x})^2} \qquad b'=\frac{\Sigma(x'-\bar{x}')(y'-\bar{y}')}{\Sigma(x'-\bar{x}')^2}$$

and the variance of their difference will be

$$\frac{\sigma^2}{\Sigma(x-\bar{x})^2}+\frac{\sigma^2}{\Sigma(x'-\bar{x}')^2}$$

As in testing the difference between means it is first necessary to test whether the estimated variances in the two groups are significantly different, and if not to obtain a pooled estimate of variance. For example, suppose two regressions based upon 20 and 30 pairs of observations respectively have the following sums of squares and products:

$$\Sigma(x-\bar{x})^2=10 \qquad \Sigma(x-\bar{x})(y-\bar{y})=4 \qquad \Sigma(y-\bar{y})^2=10$$

$$\Sigma(x'-\bar{x}')^2=20 \qquad \Sigma(x'-\bar{x}')(y'-\bar{y}')=5 \qquad \Sigma(y'-\bar{y}')^2=20$$

The regression coefficients for these two sets of observations are 0·40 and 0·25 respectively and the analyses of variance are:

	For y			*For y′*		
	D.f.	*S.s.*	*M.s.*	*D.f.*	*S.s.*	*M.s.*
Regression	1	1·60	—	1	1·25	—
Residual variation	18	8·40	0·467	28	18·75	0·670
Total	19	10·00		29	20·00	

The ratio 1·44 of the residual mean squares can be tested by the variance-ratio table as in Chapter 3. Since it is not significant a pooled estimate of variance may be obtained: $(8{\cdot}40+18{\cdot}75)/(18+28)=0{\cdot}590$. The standard error of the difference between the regression coefficients is now given by

$$\sqrt{\left(\frac{0{\cdot}590}{10}+\frac{0{\cdot}590}{20}\right)}=\pm 0{\cdot}297$$

To establish whether the difference 0·15 between the regression coefficients is significant, it is best to test $0{\cdot}15/0{\cdot}297=0{\cdot}51$ by the t table (*Table V*) with 46 degrees of freedom. It is obvious that the difference is not significant.

The difference between any number of regression coefficients may be tested in this manner but it is convenient to have one overall test for the differences between a series of regressions. This can be carried out using the analysis of variance.

If the sums of squares and products given above are combined, then:

$$\Sigma (x-\bar{x})^2+\Sigma (x'-\bar{x}')^2=30$$

$$\Sigma (x-\bar{x})(y-\bar{y})+\Sigma (x'-\bar{x}')(y'-\bar{y}')=9$$

$$\Sigma (y-\bar{y})^2+\Sigma (y'-\bar{y}')^2=30$$

and if a joint regression is carried out the resulting regression coefficient is $9/30=0{\cdot}3$ and the analysis of variance as follows:

For y and y′

	D.f.	*S.s.*
Joint regression	1	2·70
Residual variation	47	27·30
Total	48	30·00

This joint regression removes a sum of squares of 2·70 compared with the total $1{\cdot}60+1{\cdot}25=2{\cdot}85$ removed by the separate regressions (with two degrees of freedom). The difference between these two values evidently tests the difference between the regressions. The following analysis of variance may therefore be constructed:

For y and y′

	D.f.	*S.s.*	*M.s.*	*V.r.*
Joint regression	1	2·70	2·70	4·58
Difference between regressions	1	0·15	0·15	0·26
Sum of separate regressions	2	2·85	—	—
Residual variation	46	27·15	0·590	
Total	48	30·00		

Note that the joint regression is significant but the difference between the regression coefficients is not. When testing the difference between two groups this test is, of course, equivalent to the test given above* but this

* As usual here the variance ratio is the square of t *i.e.* $0{\cdot}26=(0{\cdot}51)^2$.

alternative approach allows a test to be made of the difference between several regression coefficients simultaneously.

This approach, it might be observed, bears a very close resemblance to the test for the difference between several means and is derived in a similar manner.

7.2 *Comparison of regressions*—Comparison of regression coefficients does not complete the possible comparisons between two regressions. It is possible to have two or more parallel lines with the same slope *i.e.* with the same regression coefficients, but which are separate and distinct. It is then necessary to test whether the distance between the lines is significantly different from zero and it is this problem which is considered in this section.

It is not of course possible to test the distance between lines unless they are parallel. Thus in testing the difference between regressions it is necessary first to test (or assume) that the regression coefficients are not significantly different. The comparison of two regressions is consequently carried out in three steps:

1 The residual mean squares in the regressions are compared using the variance-ratio test to see whether the variations about the lines are comparable. In making this test it should be remembered that if the larger variance is divided by the smaller, the percentage derived from the variance-ratio table should be multiplied by two. If the two residual mean squares are comparable they may then be pooled.

2 The regression coefficients should then be compared using the t or variance-ratio test in the manner described above. If they are not significantly different the joint regression coefficient may be used.

3 The distance between the lines should be tested to see if it is significantly different from zero. The equations of the lines may be denoted by

$$y=\bar{y}+b\,(x-\bar{x})$$

and

$$y'=\bar{y}'+b\,(x-\bar{x}')$$

where $\bar{x}$, $\bar{y}$ and $\bar{x}'$, $\bar{y}'$ are the means in the two regressions and b is the joint regression coefficient. The distance between the two lines is then

$$y-y'=\bar{y}-\bar{y}'-b\,(\bar{x}-\bar{x}')$$

and its variance will consist of three portions:

i the variance of $\bar{y}=\sigma^2/n$

ii the variance of $\bar{y}'=\sigma^2/m$

iii the variance of $b\,(\bar{x}-\bar{x}')=\sigma^2\,(\bar{x}-\bar{x}')^2/\left(\begin{matrix}\text{joint sum of squares of}\\ \text{the independent variables}\end{matrix}\right)$

$$= \frac{\sigma^2 (\bar{x}-\bar{x}')^2}{\Sigma (x-\bar{x})^2 + \Sigma (x'-\bar{x}')^2}$$

The variance of the distance between the lines is thus

$$\frac{\sigma^2}{n} + \frac{\sigma^2}{m} + \frac{\sigma^2 (\bar{x}-\bar{x}')^2}{\Sigma (x-\bar{x})^2 + \Sigma (x'-\bar{x}')^2}$$

$$= \sigma^2 \left[\frac{1}{n} + \frac{1}{m} + \frac{(\bar{x}-\bar{x}')^2}{\Sigma (x-\bar{x})^2 + \Sigma (x'-\bar{x}')^2}\right]$$

Consequently the standard error of the distance between the two lines may be estimated and tested to see whether it is significantly different from zero.

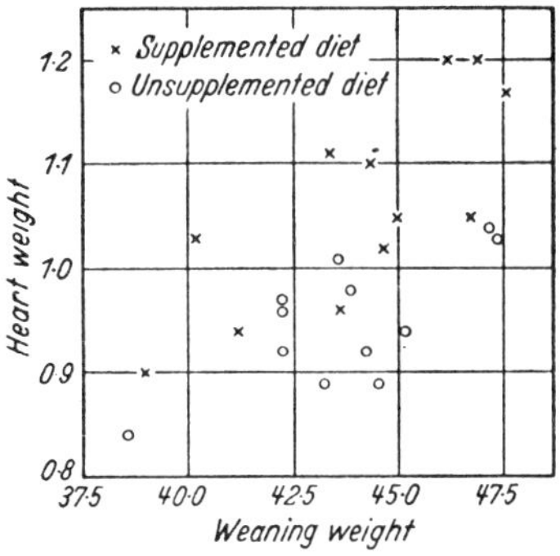

Figure 34. Graph of heart weight plotted against weaning weight for mice on two different diets

An example will serve to illustrate this whole procedure.

Figure 34 shows the dependence of heart weight h of mice at six weeks upon their weaning weight w. The analysis for the 12 mice on the supplemented diet proceeds as follows:

$$\bar{h} = 1{\cdot}061 \qquad \bar{w} = 44{\cdot}07$$

$$\Sigma (h-\bar{h})^2 = 0{\cdot}1081 \qquad \Sigma (h-\bar{h})(w-\bar{w}) = 2{\cdot}303 \qquad \Sigma (w-\bar{w})^2 = 81{\cdot}95$$

$$b = 2{\cdot}303/81{\cdot}95 = 0{\cdot}02810$$

and the regression line is

$$h = 1{\cdot}061 + 0{\cdot}02810\,(w - 44{\cdot}07)$$

The corresponding analysis of variance is:

	D.f.	*S.s.*	*M.s.*
Regression	1	0.0647	—
Residual variation	10	0.0434	0.00434
Total	11	0.1081	

For the 12 mice on the unsupplemented diet the analysis is:

$$\bar{h}' = 0{\cdot}949$$

$$\bar{w}' = 43{\cdot}70$$

$$\Sigma (h'-\bar{h}')^2 = 0{\cdot}0407 \qquad \Sigma (h'-\bar{h}')(w'-\bar{w}') = 1{\cdot}115 \qquad \Sigma (w'-\bar{w}')^2 = 62{\cdot}28$$

$$b' = 1{\cdot}115/62{\cdot}28 = 0{\cdot}01790$$

and the regression line is

$$h' = 0{\cdot}949 + 0{\cdot}01790\,(w' - 43{\cdot}70)$$

The analysis of variance here is:

	D.f.	*S.s.*	*M.s.*
Regression	1	0.0200	—
Residual variation	10	0.0207	0.00207
Total	11	0.0407	

The ratio of the residual mean squares here is 0·00434/0·00207 = 2·1 and is not significant so that they might be combined. The pooled estimate is then (0·0434 + 0·0207)/20 = 0·00320.

The variance of the difference between the two regression coefficients may now be calculated

$$0{\cdot}00320\left(\frac{1}{81{\cdot}95}+\frac{1}{62{\cdot}28}\right)=0{\cdot}0000904$$

The difference between the regression coefficients is 0·0102 and its standard error is $\sqrt{(0{\cdot}0000904)}=0{\cdot}0095$ (with 20 degrees of freedom) so that it is not significant.

Now add the sums of squares and products from the two analyses to give:

$$\text{Overall regression coefficient}=(2{\cdot}303+1{\cdot}115)/(81{\cdot}95+62{\cdot}28)$$
$$=3{\cdot}418/144{\cdot}23$$
$$=0{\cdot}02370$$

$$\text{Sum of squares due to regression}=(3{\cdot}418)^2/144{\cdot}23$$
$$=0{\cdot}0810$$

	D.f.	*S.s.*	*M.s.*
Overall regression	1	0·0810	——
Residual	21	0·0678	0·00323
Overall total	22	0·1488	

The residual sum of squares has 21 degrees of freedom since it now includes a degree of freedom corresponding to the difference between the regression coefficients. This may be removed if required (it is 0·0647 + 0·0200 − 0·0810 = 0·0037) but since the difference has already been tested and found insignificant this is unnecessary.

The two fitted regression lines are now:

$$h\ =1{\cdot}061+0{\cdot}02370\,(w\ -44{\cdot}07)$$
$$h'=0{\cdot}949+0{\cdot}02370\,(w'-43{\cdot}70)$$

The distance between these lines is

$$1{\cdot}061-0{\cdot}949-0{\cdot}02370\,(44{\cdot}07-43{\cdot}70)=0{\cdot}103$$

and its standard error is

$$\sqrt{\left[0{\cdot}00323\left\{\frac{1}{12}+\frac{1}{12}+\frac{(44{\cdot}07-43{\cdot}70)^2}{144{\cdot}23}\right\}\right]}=\pm 0{\cdot}0232$$

with 21 degrees of freedom. The value of t is thus 4·44. From *Table V* the observed distance between the lines would occur by chance less than once in a thousand times so that there is a significant difference between the regressions.

Often it is possible to assume that the residual mean square and regression coefficients are not significantly different in which case only the distance between the regression lines need be tested. The next few sections will deal with the testing of such differences.

7·3 *Concomitant observations*—The distance between two regression lines may be tested by an alternative approach. For the example of the last section the difference in mean weaning weights between the two groups is 0·37 and the difference between the mean heart weights is 0·112. The regression coefficient indicates that for each unit difference in weaning weight there is a difference of 0·02370 in heart weight, so that only $0{\cdot}37\times 0{\cdot}02370=0{\cdot}009$ of the difference in mean heart weights could be accounted for by the observed difference in weaning weight. The remainder 0·103 must be ascribed to other causes. It is this quantity which measures the distance between the regression lines.

This approach shows that testing the distance between the regression lines is equivalent to testing the difference between heart weights when the effect of possible differences in weaning weights has been removed. This removal allows the difference between the two groups to be specified more accurately. The unadjusted difference between the two groups is $0{\cdot}112 \pm 0{\cdot}0336$ as compared with the adjusted difference $0{\cdot}103 \pm 0{\cdot}0232$. The change due to the adjustment is not large but the decrease in the standard error is comparable with the effect produced by doubling the number of observations *i.e.* with a reduction of the standard error by a factor $1/\sqrt{2} = 0{\cdot}7$. Thus by eliminating the effect of differences in weaning weight the comparison between the two groups is made roughly twice as accurately.

It is now seen that the comparison of regression is the same as comparing the means of the dependent variables when the effects of changes in the independent variables are being eliminated. Usually, as a consequence of this elimination, the comparisons between the dependent variables are made more accurately. In experimentation this concept is often quite important. For instance, in the above example, the dietary comparison is improved by eliminating the effects of initial weight differences; in field experimentation the comparison of treatments of varieties might be greatly helped if some measurement reflecting initial differences in fertility can be used to eliminate these differences; or in survey work the accuracy of any comparison might be improved by 'standardizing' or eliminating uncontrollable factors, such as age, which might influence the observations. Supplementary measurements of this type which may be used to account for some of the variability are known as concomitant observations.

It is not desired to eliminate such variability if the difference which is being tested is altered as a result; the concomitant observations should not reflect any difference between the groups that are being tested. For example, if the economic states of samples of individuals from, say, Bournemouth and Brighton are being compared we should not standardize for age since the difference in ages of individuals from the two towns represents a real difference between the two towns. The effect of standardization for age *i.e.* elimination of age influences, may completely distort or nullify the comparison. In the example of the last section, since weaning weights cannot reflect dietary differences, the effects of different weaning weights may be eliminated and so the accuracy of the dietary comparison can be improved. However, the weights at death of these animals could not have been used since elimination of the effect of different weights at death would also eliminate dietary effects. In consequence, the accuracy of the comparison between the two groups would have been much reduced.

The above test of the distance between two lines attaches a standard error to the distance, but in order to be able to test simultaneously the distances between several lines *i.e.* the differences between several adjusted means, it is necessary to use the structure of the analysis of variance. The appropriate analysis will be given in the next section.

7.4 *Analysis of covariance*—To test the differences between two or more groups of measurements when they have been adjusted to eliminate the effects of other measurements, it is necessary first to carry out analyses of variance on each set of measurements. Thus, for the example of section 7.2, the analyses of variance are:

	D.f.	*Heart weights S.s.*	*Weaning weights S.s.*
Between diets	1	0·0748	0·80
Within diets	22	0·1488	144·23
Total	23	0·2236	145·03

Now corresponding to each sum of squares in these analyses there will be a sum of products of the two measurements and so, in addition, an analysis of covariance may be carried out:

	D.f.	*S.p. of heart weights and weaning weights*
Between diets	1	0·246
Within diets	22	3·418
Total	23	3·664

The total sum of squares of heart weights when the effects of differences in initial weights have been eliminated may now be calculated from

$$0{\cdot}2236 - \frac{(3{\cdot}664)^2}{145{\cdot}03} = 0{\cdot}1310$$

This has 22 degrees of freedom. In the same manner the sum of squares within diets when the effects of initial differences in weight are eliminated may be calculated from

$$0{\cdot}1488 - \frac{(3{\cdot}418)^2}{144{\cdot}23} = 0{\cdot}0678$$

This value, which was obtained previously, has 21 degrees of freedom.

The effect of differences in initial weights has been eliminated in both of these sums of squares, so that it is likewise eliminated in their difference.

This difference may therefore be used to test the dietary effect. The completed analysis of variance is thus:

	D.f.	S.s. of heart wts (weaning wts elim.)	M.s.	V.r.
Between diets	1	0·0632	0·0632	19·6
Residual variation	21	0·0678	0·00323	
Total	22	0·1310		

The difference between diets is, as previously*, highly significant and the difference between the mean heart weights may be adjusted using the regression coefficient 3·418/144·23=0·02370.

The use of the analysis of covariance in this manner allows us not only to test the differences between several groups simultaneously but also, where necessary, to eliminate the effects of other factors. For example, an analysis of covariance may be carried out for a randomized block experiment, eliminating the effects of blocks before the regressions and subsequent adjustments are carried out. The following examples will demonstrate various applications of covariance analysis.

a An experiment to compare the total bone ash weights *a* of rats raised upon two different diets gave the following analyses for the bone ash weights and weaning weights *w*:

	D.f.	S.s. for a	S.p. for a and w	S.s. for w
Diets	1	7,875	−2,175	601
Residual	107	363,267	66,909	65,958
Total	108	371,142	64,733	66,559

The residual mean square for *a* is 3,395, and the variance ratio 2·32 testing the difference between the diets is not significant. However, it is apparent that a large portion $(66{,}909)^2/65{,}958=67{,}874$ of the residual variation can be accounted for by differences in the weaning weights of the animals so that this might be eliminated leaving a residual sum of squares of 363,267 − 67,874 = 295,393. Correspondingly, removal of the effects of differences in weaning weights reduces the total sum of squares to

$$371{,}142-\frac{(64{,}733)^2}{66{,}559}=308{,}185$$

The completed analysis of variance is thus:

	D.f.	S.s. for a	M.s.	V.r.
Diets	1	12,792	12,792	4·59
Residual	106	295,393	2,787	
Total	107	308,185		

The variance ratio would occur less than once in twenty times by chance so that it may be concluded that there is evidence of a real difference between the diets.

* As usual the variance ratio obtained in the analysis of variance is the square of the value of *t* obtained directly. Thus $(4{\cdot}44)^2 = 19{\cdot}7$, the difference being due to rounding off errors.

The analysis may now be completed by adjusting the means. The unadjusted means are first calculated:

Diet	*I*	*II*
Bone ash weight	492·8	471·8
Weaning weights	210·2	215·9

Each unit difference in weaning weight makes an average difference of 66,909/65,958 = 1·014 in bone ash weight. The bone ash weights may thus be adjusted to a common mean weaning weight of, say 210 to give the following means:

Diet	*I*	*II*
Bone ash weight	492·6	465·8

The standard error of the difference may, if required, be calculated as above.

b As a second example consider the following sets of weights of chickens at two and six weeks of age. These are divided into four groups according to sex and according to the diet received by the hens.

Hens' diet	*Normal*				*Deficient*			
Sex of chicken	♂		♀		♂		♀	
Age in weeks	2	6	2	6	2	6	2	6
Weight	60	410	69	417	72	346	67	379
	68	462	78	439	75	388	69	397
	77	462	82	480	78	469	77	402
	87	478	88	499	83	406	78	398
Total	292	1,812	317	1,835	308	1,609	291	1,576

The analyses of variance of the weights at two and six weeks are:

		Two weeks		*Six weeks*	
	D.f.	*S.s.*	*M.s.*	*S.s.*	*M.s.*
Groups	3	120	40	13,542	4,514
Residual	12	756	63	14,992	1,249
Total	15	876		28,534	

At two weeks there is apparently no difference between the groups, but at six weeks the difference is just significant at the 5 per cent level. An analysis of covariance might therefore be carried out:

	D.f.	*S.p.*
Groups	3	431
Residual	12	2,292
Total	15	2,723

The weight at two weeks accounts for a significant portion $(2{,}292)^2/756 = 6{,}949$ of the residual variation so that the variation due to differences in initial weight may be removed. This gives the final analysis of variance:

	D.f.	*S.s.*	*M.s.*	*V.r.*
Groups	3	12,027	4,009	5·48
Residual	11	8,043	731	
Total	14	20,070		

The unadjusted means and means adjusted to a 2-week weight of 75 are:

Hens' diet	*Normal*		*Deficient*	
Sex	♂	♀	♂	♀
Weight at two weeks	73·00	79·25	77·00	72·75
Weight at six weeks	453·00	458·75	402·25	394·00
Adjusted 6-week weight	459·06	445·87	396·19	400·82

The difference between the groups can be attributed almost completely to the difference between the chickens from hens raised on the normal and deficient diets. Any particular comparison may be tested as previously. For example, if chickens from the normal hens are compared with those from the deficient hens we get:

Hens' diet	*Normal*	*Deficient*
Weight at two weeks	76·12	74·88
Weight at six weeks	455·88	398·12
Adjusted 6-week weight	452·47	398·50

The standard error of the difference between these adjusted means is

$$\sqrt{\left[731\left\{\frac{1}{8}+\frac{1}{8}+\frac{(76{\cdot}12-74{\cdot}88)^2}{756}\right\}\right]} = \pm 13{\cdot}57$$

with 11 degrees of freedom and the value of t is 3·98. This is significant at the 1 per cent level.

This example is unusual in that the concomitant observation, weight at two weeks, is observed after the treatment, hens' diet, is applied. Normally this would not be so. Here, it would be a misleading procedure to remove the effect of 2-week weight if it were not known that the deficiency in hens' diet would affect the chickens only after a few weeks.

The same result could have been achieved in the last example by using the component corresponding to the comparison of the normal and deficient diets in the analysis of variance and covariance. In the subsequent calculation the total sums of squares and products would have been replaced by the residual sums of squares and products plus this component. Similarly, if several components had to be tested they should be added to the residuals in turn and each corrected by a regression before the 'corrected' residual is subtracted.

In general, since tables of means are usually required, it is often easier to carry out a combined test, to calculate approximate standard errors for the tables of means and, subsequently, to calculate exact standard errors for large or suggestive differences.

Approximate standard errors may be calculated easily if the effect of the correction is ignored. Thus, for example, an approximate standard error for the difference between the means of the two groups tested above is given by

$$\sqrt{\left[731\left(\frac{1}{8}+\frac{1}{8}\right)\right]} = \pm 13{\cdot}52$$

The difference between this and the exact value is of no practical importance but it should be noted that the corrections here were not very large. For the original four groups, the standard error of a difference is given approximately by

$$\sqrt{\left[731\left(\frac{1}{4}+\frac{1}{4}\right)\right]} = \pm 19{\cdot}12$$

From this value it is apparent that only the difference in hens' diet is of importance in this experiment.

It must be also noted that, since a term is omitted in calculating the approximate standard error, the true value is always underestimated. In consequence, any difference that is not significant using the approximate value cannot be significant with the exact value. The converse is not true and exact estimates of the standard errors should be made for differences which are not very highly significant.

7.5 *Use of analysis of covariance in estimation of associations*—The analysis of covariance may be used to estimate the correlation between two variables or the dependence of one variable upon another when the effects of other quantities have been removed. Thus, if a regression has been carried out another variable may be introduced and tested for significance by the analysis of covariance. Alternatively, the analysis of variance may be used to eliminate differences between groups into which the data may be sorted, and the correlation tested with another variable using analysis of covariance.

The following example will demonstrate how this may be done. The data given below show the variations in numbers of deaths and mean quarterly temperatures in Scotland during the years 1943-47.

Deaths in Thousands, d

Quarter \ *Year*	1943	1944	1945	1946	1947	*Total*
First	17·7	18·0	18·7	19·7	21·2	95·3
Second	15·9	15·3	15·4	15·2	15·7	77·5
Third	14·2	14·7	13·1	13·6	13·4	69·0
Fourth	18·9	16·6	15·5	16·1	15·9	83·0
Total	66·7	64·6	62·7	64·6	66·2	324·8

Mean Temperature t°F

Quarter \ *Year*	1943	1944	1945	1946	1947	*Total*
First	42·1	40·9	40·8	40·2	34·7	198·7
Second	51·2	50·5	50·8	50·5	50·8	253·8
Third	55·5	56·6	57·9	56·4	58·7	285·1
Fourth	44·5	42·5	45·7	43·5	44·5	220·7
Total	193·3	190·5	195·2	190·6	188·7	958·3

There is a strong seasonal fluctuation in both sets of figures. However this does not necessarily mean that the numbers of deaths are influenced by temperature since both may be independently subject to seasonal fluctuations. Seasonal differences might therefore be eliminated before testing this dependence. Also to eliminate any small changes in the structure and size of the population during this period differences between years may also be eliminated before carrying out this calculation. The analyses of variance and covariance for this purpose are:

	D.f.	S.s. for d	S.p. for d and t	S.s. for t
Years	4	2·49	−2·00	6·62
Quarters	3	72·92	−243·67	860·39
Residual	12	14·84	−15·97	39·62
Total	19	90·25	−261·64	906·63

It is seen that temperature accounts for $(-15{\cdot}97)^2/39{\cdot}62=6{\cdot}44$ of the residual sum of squares and that this regression gives a variance ratio of 8·43 which is significant at the 5 per cent level. The regression coefficient −0·403 indicates that for each degree fall in mean quarterly temperature there are on average 403 more deaths in Scotland.

The analysis of covariance may now be used to test whether the differences in the quarterly numbers of deaths may be completely ascribed to temperature differences. Either the mean quarterly death rates may be corrected to comparable temperatures or the significance of differences may be tested after temperature differences have been removed. If the latter method is adopted residual + quarters is calculated first in each analysis:

	D.f.	S.s. for d	S.p. for d and t	S.s. for t
Residual + quarters	15	87·76	−259·64	900·01

The regression here accounts for a sum of squares $(-259{\cdot}64)^2/900{\cdot}01=74{\cdot}90$ so that the analysis of variance testing the differences between quarters when temperature effects have been eliminated is:

	D.f.	S.s.	M.s.	V.r.
Quarters	3	4·46	1·487	1·946
Residual	11	14·84 − 6·44 = 8·40	0·764	
Residual + quarters	14	87·76 − 74·90 = 12·86		

As large differences between quarters as these would occur more than once in ten times by pure chance so that evidently the differences between the mean quarterly deaths can be accounted for by temperature changes. If, in addition, these means are corrected to a mean quarterly temperature of 50°F the following values are obtained:

Quarter	*First*	*Second*	*Third*	*Fourth*
Adjusted mean	14·9	15·8	16·6	14·2
Approximate standard error of differences		±0·55		

The approximate standard error, here, is greatly underestimated owing to the large corrections involved. Exact estimates of the standard errors would be required for testing differences between the second and third quarters on the one hand and first and fourth quarters on the other.

7.6 *Regression of group means*—In the example of the last section the dependence of the number of deaths upon the temperature was estimated by eliminating the effects of seasons and years. However, this dependence might alternatively have been estimated using the quarterly means, but such a procedure would give an incorrect impression of the effect of temperature if there existed a difference in quarterly death rates independent of temperature. Any difference between the regression based upon the quarterly means and that based upon the residuals would thus be indicative of a real difference, independent of temperature, between the quarters.

In this example the regression coefficient based upon quarterly means is $-243{\cdot}67/860{\cdot}39 = -0{\cdot}283$. This accounts for 69·01 of the sum of squares of the quarters leaving a portion 3·91 with 2 degrees of freedom when the effect of temperature has been eliminated. This value might be compared with the 4·46 with 3 degrees of freedom obtained using the residual regression. The difference, evidently, is due to the difference between the regressions calculated from the seasonal means and from the residual. The sum of squares due to the difference between quarters may thus be broken into two parts, one of which tests the difference between the two regressions, in the following manner:

	D.f.	*S.s.*	*M.s.*
Difference between regressions	1	0·55	0·550
Quarters, adjusted by quarters regression	2	3·91	1·955
Quarters, adjusted by residual regression	3	4·46	—

Both of these components test, of course, an effect of quarters.

An alternative, and useful, method of regarding this analysis is provided by considering the adjusted means of the last section together with the temperature means:

Quarter	*First*	*Second*	*Third*	*Fourth*
Adjusted mean No. of deaths	14·9	15·8	16·6	14·2
Mean temperature	39·7	50·8	57·0	44·1

Apparently the adjusted mean number of deaths increases with temperature, indicating that there may be some additional effect of temperature between seasons which does not occur within seasons. It is this apparent trend of the adjusted means with temperature which is tested by the single degree of freedom, 0·55. Since the residual mean square is 0·764 the effect is far from significant.

This same form of analysis may also be employed to test the difference between the regressions calculated from the variation within any set of groups and from the variation between the group means. Usually we are not specifically interested in this difference and the overall sum of squares for the adjusted group means (the 4·46 above) should be used to test the difference between groups. On some occasions, however, the changes in these adjusted means with the concomitant means are of interest and then the single degree of freedom testing this may be isolated as above.

SUMMARY OF PP 134 TO 147

The procedure for testing the difference between two or more regressions using the analysis of variance has been given. It has been demonstrated, using the analysis of covariance how it is possible to eliminate the variation ascribable to concomitant observations and hence to improve the accuracy of the comparisons that are being tested.

The use of the analysis of covariance in estimating and testing the associations between different variables after the elimination of the effects of other quantities has been illustrated. Finally, it has also been shown how the analysis of covariance may be used to compare regressions calculated within and between groups.

EXAMPLES

63 Using the data of example *62* calculate the following analyses of variance and covariance and use them to show that even when the effect of height has been eliminated, weight is still correlated with metabolic rate:

	D.f.	*S.s. for m*	*S.p. for m and w*	*S.s. for w*
Regression on h	1	1,779,972	88,746*	4,425
Residual	219	4,509,088	288,577	22,277
Total	220	6,289,060	377,323	26,702

This gives yet another approach to the partial correlation coefficient.

*Since the sums of squares due to the regressions in the two analyses of variance are

$$[\Sigma (h-\bar{h})(m-\bar{m})]^2/\Sigma (h-\bar{h})^2$$

and

$$[\Sigma (h-\bar{h})(w-\bar{w})]^2/\Sigma (h-\bar{h})^2$$

the sum of products in the analysis of covariance is correspondingly

$$\frac{[\Sigma (h-\bar{h})(m-\bar{m})]\,[\Sigma (h-\bar{h})(w-\bar{w})]}{\Sigma (h-\bar{h})^2}$$

64 For the data of example b in section 7.4 show that the differences between the regression coefficients for the four dietary groups are insignificant using the analysis of variance:

	D.f.	*S.s.*	*M.s.*	*V.r.*
Overall regression	1	6,949	6,949·0	8·56
Difference between regressions	3	1,547	515·7	0·62
Sum of individual regressions	4	8,496	——	——
Residual	8	6,496	812·0	
Total within groups	12	14,992		

65 The mean biacromial breadths b of two groups of 200 boys and 200 girls aged between 6 and 11 years were 27·32 and 27·41 cm respectively. The mean ages a of these two groups were 8·51 and 8·93 years and the corrected total sums of squares and products were:

$$\Sigma\,(b-\bar{b})^2=851{\cdot}32 \qquad \Sigma\,(b-\bar{b})\,(a-\bar{a})=407{\cdot}29 \qquad \Sigma\,(a-\bar{a})^2=447{\cdot}42$$

Find the average rate of increase in biacromial breadth per year and show that when adjusted for age differences biacromial breadth is greater for the boys than for the girls.

66 The results of a uniformity trial u were used in an analysis of covariance with the yields y of grain in the following:

	D.f.	*S.s. for y*	*S.p. for y and u*	*S.s. for u*
Blocks	3	111·2	106·5	150·4
Treatments	7	63·9	18·2	47·2
Residual	21	132·6	79·2	126·6
Total	31	317·7	203·9	324·2

Show that the differences between the treatments are not significant.

67 The following table gives the total rainfall in inches in Scotland during the period 1943-47:

Quarter \ *Year*	1943	1944	1945	1946	1947	*Total*
First	13·3	10·7	15·2	12·5	11·6	63·3
Second	11·8	11·3	11·2	7·8	14·2	56·3
Third	13·5	11·6	12·5	16·6	9·5	63·7
Fourth	14·7	19·4	12·1	13·1	13·5	72·8
Total	53·3	53·0	51·0	50·0	48·8	256·1

Show that, if differences between years and between quarters are eliminated, the correlation of the residuals with the number of deaths, given in section 7.5 is 0·110, but that if this elimination is not carried out the correlation is 0·114. Neither of these values is significant.

68 Show that the use of the single degree of freedom in the analysis of variance to test the difference between the two regression coefficients which may be derived using the residuals and the treatment components of the analysis of variance is equivalent to the use of a t test in comparing these regression coefficients.

EXTENDED DEVELOPMENT

7A.7 *Analysis of covariance with two or more sets of concomitant observations*—The analysis of covariance and its applications are not, of course, limited to a single set of concomitant observations. If two or more sets are used then multiple regressions have to be calculated instead of regressions on single variables. Consequently, analyses of covariance have to be calculated for each pair of variables. Apart from these differences the procedure in the analysis is unaltered. An example will demonstrate the method.

The following table gives the number of hours of sunshine s in Scotland during the years 1943-47:

Quarter \ Year	1943	1944	1945	1946	1947	*Total*
First	198	181	186	209	202	976
Second	473	361	480	520	396	2,230
Third	397	357	377	344	492	1,967
Fourth	142	146	150	150	158	746
Total	1,210	1,045	1,193	1,223	1,248	5,919

If an analysis of covariance is carried out to test whether the numbers of deaths in Scotland, given in section 7.5 are correlated with hours of sunshine we get:

	D.f.	*S.p. for d and s*	*S.s. for s*
Years	4	35·0	6,424
Quarters	3	−3,428·8	318,448
Residual	12	−78·3	25,143
Total	19	−3,472·1	350,015

Thus the number of hours of sunshine accounts for only $(-78{\cdot}3)^2/25{,}143 = 0{\cdot}24$ of the residual variation. This does not however rule out the possibility that when used in conjunction with the mean temperature it might contribute significantly to the analysis. To test this the analysis of covariance for temperature and sunshine is calculated:

	D.f.	*S.p. for t and s*
Years	4	1·2
Quarters	3	13,458·9
Residual	12	163·6
Total	19	13,623·7

The regression coefficients for the joint regression may now be determined from the equations:

$$39{\cdot}62b_1 + 163{\cdot}6b_2 = -15{\cdot}97$$

$$163{\cdot}6b_1 + 25{,}143b_2 = -78{\cdot}3$$

from which $b_1 = -0{\cdot}4010$, $b_2 = -0{\cdot}00051$. This accounts for a sum of squares 6·44 in the residual, which is exactly the same as was accounted for by temperature alone. Evidently what small correlation there is between the numbers of deaths and the hours of sunshine can be ascribed to the effect of temperature.

Since the number of hours of sunshine is not significant it should not be included in further calculations and the analysis would be completed as previously. However, if the contribution from this variable had been significant further regressions with the two variables might have been used in place of the regressions with one variate. The previous tests could then have been repeated correcting for the effects of both mean temperature and hours of sunshine.

7A.8 *Dummy variates*—Often it is possible to classify observations according to some ordered system which, nevertheless, is incapable of exact measurement. Thus a response might be classified as very good, good, fair, poor or very poor or according to a 5-point scale 4, 3, 2, 1 or 0. The latter use of numbers to denote the classifications implies an equidistant ordering which may not be exactly realized in practice but which is useful for the purposes of analysis. Thus, in *Figure* 25 on p 105, the classification of degree of concentration according to a 5-point scale allows the average increase in intelligence quotient for different degrees of concentration to be estimated using the regression technique. In the same manner, it is often useful when data are classified into two groups to assign the value 0 to one group and 1 to the other. The average increase here is simply the difference between the groups, but it is convenient to be able to test this using the regression technique. Such assigned values are called dummy variates since they do not represent any actual measurement but only the ordered classification of the data into two or more groups.

Two examples will demonstrate the use of dummy variates in experimentation:

						Total
E 39·32	*A* 33·71	*F* 41·49	*C* 30·68	*B* 40·41	*D* 38·46	224·07
C 35·22	*E* 28·09	*B* 31·55	*F* 19·66	*D* 23·77	*A* 15·56	153·85
F 27·01	*C* 41·49	*D* 28·95	*B* 28·09	*A* —	*E* 28·95	154·49
B 41·92	*F* 30·03	*A* 15·56	*D* 31·11	*E* 30·03	*C* 31·11	179·76
D 39·32	*B* 42·57	*E* 37·81	*A* 21·82	*C* 21·82	*F* 25·93	189·27
A 18·58	*D* 42·57	*C* 38·89	*E* 29·60	*F* 32·19	*B* 33·27	195·10
Total 201·37	218·46	194·25	160·96	148·22	173·28	1,096·54

Overall mean = 30·459

Treatment totals					
A	*B*	*C*	*D*	*E*	*F*
105·23	217·81	199·21	204·18	193·80	176·31

a The above table gives the hay yields in cwt/acre testing the effects of 6 different fertilizer dressings in a 6 × 6 Latin square. Unfortunately an accident destroyed the yield of one of the plots so that, although this plot gave a yield of hay, its magnitude was not known. This situation is not uncommon in experimentation and it presents a difficulty in the analysis.

If the missing plot yield is ignored and the experiment analysed as if the yield had been zero, the resulting analysis of variance is:

	D.f.	*S.s.*	*M.s.*	*V.r.*
Rows	5	590·9	118·2	—
Columns	5	585·2	117·0	—
Treatments	5	1,355·4	271·1	10·8
Residual	20	500·7	25·03	
Total	35	3,032·2		

However, this analysis is incorrect in that the difference between the variates is increased and changed by the missing observation, while the unaccountable variation is also swollen.

To overcome this difficulty a dummy variate is introduced taking a value one for the missing observation and zero for all other observations. If an analysis of covariance is now carried out on the dummy variate it is possible to determine how much of the residual variation is due to the missing observation and, consequently, to correct and test the differences between treatments. This analysis is easily carried out since the majority of the values taken by the dummy variate are zero. Thus the sum of products for rows is $154{\cdot}49/6-30{\cdot}459=-4{\cdot}711$, with corresponding values for the columns and treatments sums of products. The sums of squares are equally easily calculated. Thus the total sum of squares is $1^2-1^2/36=0{\cdot}97222$ and the sum of squares for rows is $1^2/6-1^2/36=0{\cdot}13889$. The analyses of variance and covariance for the dummy variate are thus:

	D.f.	*S.p. for dummy variate*	*S.s. for dummy variate*
Rows	5	−4·711	0·13889
Columns	5	−5·756	0·13889
Treatments	5	−12·921	0·13889
Residual	20	−7·071	0·55555
Total	35	−30·459	0·97222

The regression coefficient is $b=-7{\cdot}071/0{\cdot}55555=-12{\cdot}73$ and this gives the amount by which the missing plot must be adjusted. Thus the estimated yield of the missing plot is 12·73 cwt/acre and the estimated mean yield for treatment A is $(105{\cdot}23+12{\cdot}73)/6=19{\cdot}66$. The regression accounts for $(-7{\cdot}071)^2/0{\cdot}55555=90{\cdot}0$, of the residual sum of squares leaving 410·7 with 19 degrees of freedom. To test the significance of the difference between treatments it is necessary to add the treatments sum of squares and products to the residual giving:

	D.f.	*S.s.*	*S.p. for dummy variate*	*S.s. for dummy variate*
Treatments+residual	25	1,856·1	−19·992	0·69444

When the correction for the missing plot is carried out, the treatments+residual sum of squares is reduced to $1{,}856{\cdot}1-(-19{\cdot}992)^2/(0{\cdot}69444)=1{,}282{\cdot}6$ and the analysis of variance testing the effect of treatments is:

	D.f.	*S.s.*	*M.s.*	*V.r.*
Treatments	5	871·9	174·38	8·07
Residual	19	410·7	21·62	
Treatments+residual	24	1,282·6		

The difference between treatments is highly significant so that a table of treatment means may be constructed:

Treatment means					
A	*B*	*C*	*D*	*E*	*F*
19·66	36·30	33·20	34·03	32·30	29·38

The standard errors of the differences between means are calculated as usual if we note that the mean A is corrected by $\frac{1}{6}b$. This gives:

Standard error of the difference between A and any other treatment

$$=\sqrt{\left[21{\cdot}62\left\{\frac{1}{6}+\frac{1}{6}+\frac{(0{\cdot}1667)^2}{0{\cdot}55555}\right\}\right]}$$

$$=\pm 2{\cdot}88$$

Standard error of any other difference

$$=\sqrt{\left[21{\cdot}62\left\{\frac{1}{6}+\frac{1}{6}\right\}\right]}$$

$$=\pm 2{\cdot}68$$

Obviously treatment A has given significantly lower yields than the other treatments. The only other significant comparison is between treatments B and F.

b

				Totals
Blocks	*C* 15·11	*G* 15·33	*D* 14·79	
	B 11·21	*I* 15·68	*A* 23·44	136·59
	H 13·60	*E* 12·41	*F* 15·02	
	I 15·40	*D* 11·94	*H* 22·25	
	G 11·16	*F* 13·16	*C* 19·27	139·75
	B 10·67	*A* 16·29	*E* 19·61	
	F 9·30	*H* 13·90	*G* 13·52	
	B 10·84	*E* 15·32	*I* 23·06	114·52
	A 9·18	*C* 7·22	*D* 12·18	
Column Totals	106·47	121·25	163·14	390·86

Treatment	*Totals*
A	48·91
B	32·72
C	41·60
D	38·91
E	47·34
F	37·48
G	40·01
H	49·75
I	54·14
Total	390·86

The previous table gives the lay-out of a randomized block experiment to test the effects of nine treatment combinations of lime and phosphate on the grain yield of an oats crop. Thus *A, B, C; D, E, F;* and *G, H, I* each represent different lime treatments, while *A, D, G; B, E, H;* and *C, F, I* each represent different phosphate treatments. The analysis of variance for the yields y of this experiment is:

	D.f.	*S.s. for y*	*M.s.*	*V.r.*
Lime	2	30·90	15·45	0·89
Phosphate	2	1·65	0·82	0·05
Lime × phosphate	4	95·97	23·99	1·38
Treatments	8	128·52	16·07	—
Blocks	2	41·99	—	—
Residual	16	277·71	17·36	
Total	26	448·22		

The residual mean square in this experiment is very large; the coefficient of variation is 29 per cent compared with the usual 10 to 15 per cent for agricultural experiments. An examination of the plan shows that a large fertility trend has occurred across the field; the third column giving very high yields compared with the first. It is this trend which accounts for the size of the residual mean square.

The sum of squares due to columns cannot be directly removed since it is not orthogonal to the treatment sum of squares. To achieve orthogonality each treatment would have to occur once in each column, but treatment *B* occurs three times in the first column and not at all in the second and third columns. Two dummy variates might be introduced: d_1 which takes the value 1 for the first column and zero elsewhere and d_2 which takes the value 1 for the second column and zero elsewhere. The dummy variates d_1 and d_2 thus measure the differences between the first two columns and the third column. The analyses of variance and covariance are then:

	D.f.	*S.p. for y and d_1*	*S.p. for y and d_2*	*S.s. for d_1*	*S.p. for d_1 and d_2*	*S.s. for d_2*
Lime	2	−0·11	0·06	0·89	−0·44	0·22
Phosphate	2	0·22	0·00	0·22	0·00	0·00
Lime × phosphate	4	− 7.05	4·81	0·89	−0·56	0·45
Treatments	8	−6·94	4·87	2·00	−1·00	0·67
Blocks	2	0·00	0·00	0·00	0·00	0·00
Residual	16	−16·88	−13·91	4·00	−2·00	5·33
Total	26	−23·82	−9·04	6·00	−3·00	6·00

and the regression equations are:

$$4{\cdot}00b_1 - 2{\cdot}00b_2 = -16{\cdot}88$$

$$-2{\cdot}00b_1 + 5{\cdot}33b_2 = -13{\cdot}91$$

from which we get:

$$b_1 = -\ 6{\cdot}801 \qquad b_2 = -\ 5{\cdot}162$$

Thus the plot yields in the first two columns are on average 6·80 and 5·16 less than the plot yields in the third column. The reduction in the residual sum of squares due to this regression is 186·60 and is highly significant. The treatment means may now be adjusted and any particular comparisons tested using the covariance technique. Thus to test the differences between the lime treatments, the lime sums of squares and products must be added to the residuals, and the regression equations to be solved are then:

$$4{\cdot}89b_1 - 2{\cdot}44b_2 = -16{\cdot}99$$

$$-2{\cdot}44b_1 + 5{\cdot}55b_2 = -13{\cdot}85$$

i.e.

$$b_1 = -\ 6{\cdot}046 \qquad b_2 = -\ 5{\cdot}154$$

This regression accounts for a sum of squares 174·10 so that the analysis of variance testing the lime treatments is:

	D.f.	*S.s.*	*M.s.*	*V.r.*
Lime	2	43·40	21·70	3·33
Residual	14	277·71 − 186·60 = 91·11	6·508	
Lime+residual	16	308·61 − 174·10 = 134·51		

This variance ratio would occur slightly more than once in twenty times by pure chance so that we could not conclude that the difference between the lime treatments is significant.

7A.9 *Non-orthogonality*—When two effects are not orthogonal they cannot be tested simultaneously by the usual analysis of variance approach since there is a danger that one effect may be reflected in the other. Example *b* of the previous section provides an illustration of two effects which are not orthogonal and it also demonstrates how their significance might be tested. As a further example consider the following figures of daily calorie consumptions in families of five:

	Town		*Country*		*Total*
	Working	*Unemployed*	*Working*	*Unemployed*	
Calorie consumption	11·9	12·1	15·4	11·8	
	12·2	11·3	16·0	16·2	
	15·7	13·4	14·2	13·2	
	14·0	14·1	16·0	15·1	
	13·3	13·7	17·7		
	11·4	11·8	15·4		
	13·4	12·1	15·4		
	17·1		16·5		
	15·1		19·6		
			20·5		
			19·8		
Total	124·1	88·5	186·5	56·3	455·4
Mean	13·8	12·6	17·0	14·0	14·6

If the mean consumption of unemployed is compared with that of working families this reflects in part a difference between town and country since there is a higher proportion of unemployed in the town. In the same manner, if the mean calorie consumption is compared between town and country this reflects in part the difference between unemployed and working families. The two effects are not orthogonal. In consequence, both components cannot be separated in the analysis of variance since they will not be independent.

Here again the analysis of covariance may be used. If a dummy variate

d is used, which takes a value 1 for the unemployed families and 0 for the others, an analysis of covariance may be calculated:

	D.f.	*S.s. for c*	*S.p. for c and d*	*S.s. for d*
Town versus country	1	65·08	−3·835	0·2259
Residual	29	121·57	−12·959	6·8709
Total	30	186·65	−16·794	7·0968

The regression coefficient b is −1·886 and this gives the average amount less than the working families received by the unemployed families. This regression accounts for 24·44 of the residual variability and is highly significant. To test whether the comparison between town and country is still significant when the effect of unemployment is removed, the analysis must be completed as usual:

	D.f.	*S.s.*	*M.s.*	*V.r.*
Town versus country	1	49·78	49·78	14·3
Residual	28	97·13	3·469	
Total	29	146·91		

Mean difference between country and town $= 16{\cdot}19 - 13{\cdot}29$
$= 2{\cdot}90$

Mean difference between proportions unemployed in country and town
$= 0{\cdot}267 - 0{\cdot}438$
$= -0{\cdot}171$

Mean difference between country and town, adjusted for differences in unemployment
$= 2{\cdot}90 + 0{\cdot}171 \times 1{\cdot}886$
$= 3{\cdot}22$

Standard error of adjusted difference

$$= \sqrt{\left[3{\cdot}469\left\{\frac{1}{16}+\frac{1}{15}+\frac{(0{\cdot}171)^2}{6{\cdot}8709}\right\}\right]}$$
$$= \pm 0{\cdot}680$$

Mean difference between working and unemployed adjusted for differences in locality
$= 1{\cdot}886$

Standard error of adjusted difference $= \sqrt{\left(\frac{3{\cdot}469}{6{\cdot}8709}\right)}$
$= \pm 0{\cdot}711$

Here the effect of the non-orthogonality is not very large, but a greater inequality in the proportions of unemployed families in town and country might have produced an appreciable effect.

Two points should be noted. First, if the comparisons are non-orthogonal and each has several degrees of freedom then several dummy variates are needed. Usually it will be easier to adjust for the comparison with the lower number of degrees of freedom. Secondly, this type of adjustment using dummy variates assumes that the interactions are zero. For example, in adjusting for the effect of unemployment in this example it was assumed that it has equal effects in town and country. If this is not so, the comparison between town and country is different for the working and unemployed families. The correct procedure here is to use the means of each group in making any further comparisons. How this may be done will be considered in the next section. It should, however, be noted that it is now easy to test whether the interaction is significant.

The sum of squares due to the four groups is

$$\frac{(124{\cdot}1)^2}{9}+\frac{(88{\cdot}5)^2}{7}+\frac{(186{\cdot}5)^2}{11}+\frac{(56{\cdot}3)^2}{4}-\frac{(455{\cdot}4)^2}{31}=94{\cdot}56$$

Of this, the comparison of town with country accounts for a sum of squares 65·08, and the comparison of unemployed with working families accounts for an additional sum of squares of 24·44. The remainder 94·56 − 65·08 − 24·44 = 5·04 must be attributed to the interaction of these comparisons. This may be tested using the analysis of variance:

	D.f.	*S.s.*	*M.s.*	*V.r.*
Town versus country	1	65·08	65·08	19·08
Extra variation due to comparison unemployed versus workers	1	24·44	24·44	7·17
Interaction	1	5·04	5·04	1·48
Groups	3	94·56	—	—
Residual	27	92·09	3·411	
Total	30	186·65		

The interaction, here, is insignificant and no further analysis is necessary.

7A.10 *Non-orthogonal comparisons when interaction exists*—It has been pointed out in the previous section that if interaction exists the group means have to be considered. The existence of interaction implies, in fact, the existence of the main effects. Thus, if there had been an interaction in the example of the last section this would imply that the difference between working and unemployed families was not the same in town and country and *a fortiori* that a difference between working and unemployed families exists. There would thus seem to be little point in testing the main effects if interaction exists. The average effect may however be considered to determine whether it is significantly different from zero. Alternatively, the comparisons which have the smallest errors may be determined.

As an example consider the data given in the following table. This gives the lengths of *Plantago maritima* found in two habitats and classified according to two growth forms:

Habitat	*Length on cliff cm*		*Length on salt marsh cm*	
Growth habit	*1*	*2*	*1*	*2*
	11·0	13·0	28·0	45·0
	13·0	12·0	23·0	46·0
	15·0	15·0	28·0	33·0
	10·0	13·0	17·0	28·0
	10·0	14·0	25·0	33·0
	22·0	12·0	15·0	32·0
	12·0	15·0	22·0	
	11·5	18·0	19·0	
	6·5	15·0	28·0	
		17·0	27·0	
		10·5	26·0	
			17·0	
			25·0	
			17·0	
			17·5	
Total	111·0	154·5	334·5	217·0
Mean	12·33	14·05	22·30	36·17

The means indicate that there is probably a significant interaction between growth habit and habitat. The mean difference between salt marsh and cliff for growth form *1* is 9·97 as compared with 22·12 for growth form *2*. The interaction might therefore be estimated as $\frac{1}{2}(22{\cdot}12-9{\cdot}97)=6{\cdot}08$. To estimate the standard error of this value an analysis of variance using the four groups is calculated:

	D.f.	*S.s.*	*M.s.*	*V.r.*
Groups	3	2,566·32	855·44	39·7
Residual	37	797·46	21·553	
Total	40	3,363·78		

The standard error of the estimated interaction is thus

$$\frac{1}{2}\sqrt{\left[21{\cdot}553\left\{\frac{1}{9}+\frac{1}{11}+\frac{1}{15}+\frac{1}{6}\right\}\right]}=\pm 1{\cdot}532$$

The interaction is therefore nearly four times its standard error and is consequently highly significant. The average difference between the salt marsh and cliff is $\frac{1}{2}(22{\cdot}12+9{\cdot}97)=16{\cdot}04$ and the average difference between growth habits *1* and *2* is $\frac{1}{2}(1{\cdot}72+13{\cdot}87)=7{\cdot}80$. The standard errors of both of these values is also $\pm 1{\cdot}532$.

Alternatively these averages may be weighted so as to get as small a standard error as possible. The variance of the difference 22·12 is

$$21{\cdot}553\left(\frac{1}{6}+\frac{1}{11}\right)=21{\cdot}553\times\frac{17}{66}$$

and the variance of the difference 9·97 is

$$21{\cdot}553\left(\frac{1}{15}+\frac{1}{9}\right)=21{\cdot}553\times\frac{8}{45}$$

Using the method of weighting employed in section 3A.11 the weighted combination

$$\left(\frac{66}{17}\times 22{\cdot}12+\frac{45}{8}\times 9{\cdot}97\right)\Big/\left(\frac{66}{17}+\frac{45}{8}\right)=14{\cdot}93$$

has a standard error

$$\sqrt{\left[21{\cdot}553\Big/\left(\frac{66}{17}+\frac{45}{8}\right)\right]}=\pm 1{\cdot}506$$

The difference between this value and that obtained previously is not very large since the numbers in each group are not very different. If, however, the numbers had differed greatly, the standard error by this approach might have been appreciably less than that obtained by the above approach.

It might be asked why the standard error should be minimized. The answer to this is that if the standard error is minimized the observations are used to their fullest extent in estimating the effect and, in consequence, the value which is obtained is more representative of the difference than any other estimate. Thus, in this example, if pairs of plants are selected at random and the average difference calculated between pairs which have the same growth habit, but arise from different habitats, the resulting value is 16·48. This assumes that the original sample was, in fact, random or representative of the areas sampled but, with this assumption, it can be seen that the latter method of estimating the mean difference between habitats is at least as meaningful as the former.

The general method of combining a series of differences proceeds along the same lines as those above. However the amount of work and possible wastage of data involved with non-orthogonal data make orthogonality a desirable property in experimentation and where some control can be exercised over the observations that are taken the desirability of orthogonality should always be borne in mind.

7A.11 *Discriminant functions*—In using dummy variates in the calculation of regressions and analyses of covariance these have so far been used as the independent variables. Since the error in the dependent variable is considered in estimating the regression the use of a dummy variate in this manner is quite valid. If, however, the dummy variate is used as a dependent variable, the use of regression methods might be criticized on the ground that it is no longer subject to error or unaccountable variation. Fortunately, provided the independent variables are subject to error or unaccountable variation, the same effect is achieved and the regression method may be used. If, however, both dependent and independent variables are dummy then the regression or covariance method is no longer valid and alternative tests, such as chi-squared, should be used. Even so these methods are still approximately valid if sufficient observations are taken. See example 72 for an illustration of this approximate validity.

To demonstrate the use of a dummy variate as the dependent variable in a regression, consider the data plotted in *Figure 35*. This shows the reactions of two groups of mice to single and double doses of a drug. These reactions were measured by two observations: the length of time t before the dose ceased to have any effect (actually the logarithm of the number of days) and the logarithm of the weight increase w during this time. The difference between the values of t for the two doses was $0{\cdot}184 \pm 0{\cdot}088$ while the difference between the values of w was $0{\cdot}225 \pm 0{\cdot}116$. Neither of these quite attains the 5 per cent level of significance (with 17 degrees of freedom) but both are highly suggestive.

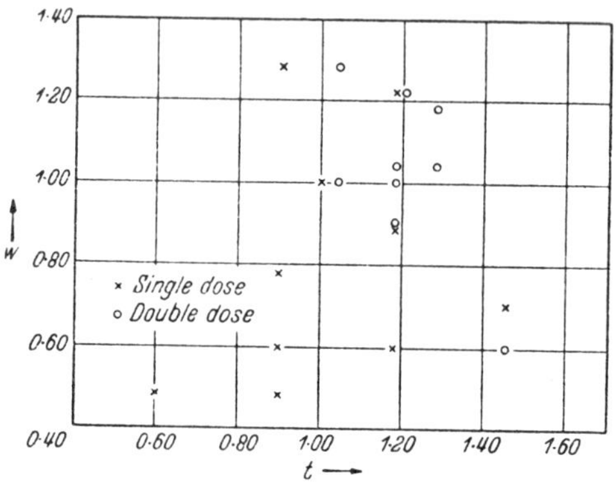

Figure 35. Plot of weight increases against response times

It might now be asked whether the use of both measurements simultaneously would lead to a better differentiation between the groups or whether a combination of the two measurements would be more sensitive to the change in the dose. To determine this, a dummy variate d is employed which takes a value 1 for the double dose and 0 for the single dose. A regression of d on t and w will then give the combination of these measurements which is most sensitive to changes in the dose level. The regression coefficients, here, may be calculated from:

$$0{\cdot}7846\, b_t + 0{\cdot}2170\, b_w = 0{\cdot}8732$$

$$0{\cdot}2170\, b_t + 1{\cdot}3289\, b_w = 1{\cdot}0658$$

where

$$\Sigma(t-\bar{t})^2=0{\cdot}7846 \quad \Sigma(t-\bar{t})(w-\bar{w})=0{\cdot}2170 \quad \Sigma(w-\bar{w})^2=1{\cdot}3289$$

$$\Sigma(t-\bar{t})(d-\bar{d})=0{\cdot}8732 \quad \Sigma(w-\bar{w})(d-\bar{d})=1{\cdot}0658$$

These equations give:

$$b_t=0{\cdot}9333 \quad b_w=0{\cdot}6496$$

and $D=0{\cdot}9333\,t+0{\cdot}6496\,w$ is thus the estimate of the most sensitive linear combination of t and w to changes in the dose. The constant in the regression equation may be ignored since this does not alter the difference between the groups. This combination is known as a discriminant function since it discriminates between the two groups receiving the single and double doses. The mean value of D for all the animals is 1·625.

Figure 36 shows the straight lines:

$$0{\cdot}9333\,t+0{\cdot}6496\,w=1{\cdot}625$$

$$0{\cdot}9333\,t+0{\cdot}6496\,w=1{\cdot}474$$

$$0{\cdot}9333\,t+0{\cdot}6496\,w=1{\cdot}792$$

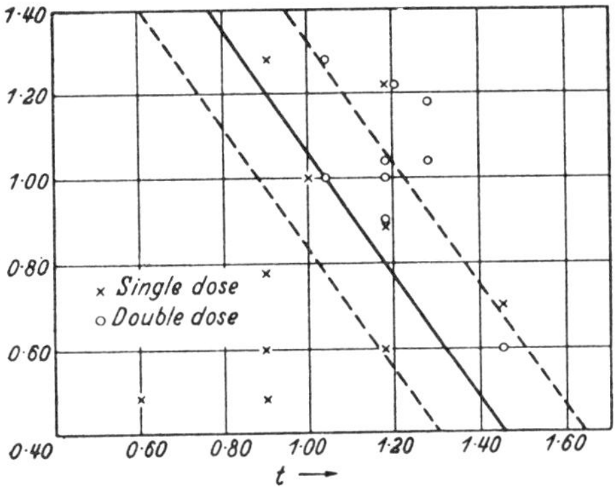

Figure 36. Discriminant function fitted to data of Figure 35

Points above the first line represent a greater than average response while those below represent a less than average response. It is seen that the double dose animals tend to lie above the line and the single dose animals below the line.

The average values of D for the two groups are 1·474 and 1·792. The other two lines which are shown dotted in *Figure 36* therefore indicate the average responses in the two groups.

To test the significance of the difference between the two groups, the analysis of variance testing the regression should be constructed. The sum of squares due to the regression is $0{\cdot}9333 \times 0{\cdot}8732 + 0{\cdot}6496 \times 1{\cdot}0658 = 1{\cdot}5073$ and the analysis of variance is:

	D.f.	*S.s.*	*M.s.*	*V.r.*
Regression	2	1·5073	0·7536	3·73
Residual	16	3·2295	0·2018	
Total	18	4·7368		

This variance ratio is significant at the 5 per cent level so that it may be concluded that the difference between the groups is real.

The significance of any particular variable may be tested in the same manner as for a regression. Here, it is of some interest to test whether the coefficients of the discriminant function are significantly different. If not, then $t+w$ might be used *i.e.* the sum of the logarithms of time and weight increases, instead of the more complicated linear function. This can be tested by carrying out a regression of d on $t+w$ which, in fact, accounts for a sum of squares of 1·4758. The difference $1{\cdot}5073-1{\cdot}4758=0{\cdot}0315$, with one degree of freedom, tests whether the regression coefficients are significantly different. Evidently, the simple combination $t+w$ is almost equally as good as the more complicated discriminant function.

This technique of deciding what measurements or combinations of measurements are most sensitive to treatment differences is not restricted to two groups only. If there are several groups and if a dummy variate can be assigned giving their relative order (as in *Figure* 25) the regression technique may be used as above to estimate the discriminant function. If a dummy variate cannot be assigned, then it is necessary to carry out a more complicated analysis to estimate their relative order at the same time. The form of this analysis is too lengthy to be described here. The interested reader should refer to the works of R. A. FISHER for a detailed account of these methods.

SUMMARY OF PP 148 TO 160

Application of the analysis of covariance with two or more sets of concomitant observations has been demonstrated.

It has been shown how, using a dummy variate, it is possible to correct for non-orthogonality and, in particular, for missing observations. The appropriate tests and estimates have been considered when the data are non-orthogonal and interaction also exists.

Finally, the method of estimating what variables or combinations of variables are most sensitive to group differences has been demonstrated.

EXAMPLES

69 The kidney weights of rats maintained on four different diets are given in the following table:

Sex\Diet	*I*	*II*	*III*	*IV*
♀	0·347 0·468	0·592 0·378 0·469	0·489 0·580 0·347	0·632 0·586 0·618
♂	0·704	0·633	0·351 0·676	0·641 0·787 0·607
Total	1·519	2·072	2·443	3·871

The analysis of variance testing the difference between the diets is:

	D.f.	*S.s.*	*M.s.*	*V.r.*
Diets	3	0·0830	0·02767	1·80
Residual	14	0·2151	0·01536	
Total	17	0·2981		

Using a dummy variable taking a value 1 for the males and 0 for the females, show that the average difference between the sexes is 0·112 and that this accounts for an additional sum of squares of 0·0518. Prove that this value is not significant.

Show also that the sum of squares ascribable to the interaction between sex and diets is 0·0331 and that it is not significant.

70 The following table gives the yields in cwt/acre of turnip yields in randomized blocks. The experiment, which was of factorial type testing two levels of phosphate and three methods of application of dung, had two missing plot yields indicated by dashes:

Dung dressing	*No dung*		*5 ton/acre applied separately*		*5 ton/acre mixed with phosphate*		
Block \\ *Phosphate dressing per acre*	*40 lb*	*80 lb*	*40 lb*	*80 lb*	*40 lb*	*80 lb*	*Total*
1	419	415	434	487	492	470	2,717
2	420	431	475	490	498	489	2,803
3	439	431	451	—	448	517	2,286
4	341	459	474	485	476	481	2,716
5	380	—	408	418	480	507	2,193
Total	1,999	1,736	2,242	1,880	2,394	2,464	12,715

Show that the estimates of yields for the missing plots are 479·3 in block *3* and 409·3 in block *5*. Show also that the effects of phosphate and dung dressings are significant but that their interaction is negligible.

Construct the following tables of means:

Dressing	*No dung*	*Dung separately*	*Dung mixed*
Mean	414·4	460·1	485·8
Approximate standard error of differences		± 12·34	

Dressing	*40 lb phosphate*	*80 lb phosphate*
Mean	442·3	464·6
Approximate standard error of differences	± 10·08	

Here instead of using the total number of plots on each treatment in calculating the approximate standard error it is preferable to use the actual number of plot yields excluding the missing plots. Thus a better value for the standard error of the difference between the two phosphate treatments is given by

$$\sqrt{\left[761\left(\frac{1}{15}+\frac{1}{13}\right)\right]} = \pm 10{\cdot}45$$

The value of t for 18 degrees of freedom which is exceeded by pure chance once in twenty times is 2·101 so that the 5 per cent level of significance for the difference is 21·98. This is just exceeded.

71 A series of observations on the lengths and heights of *Plantago maritima* gave the following figures:

Growth habit	1		3	
	Length	*Height*	*Length*	*Height*
	28·0	8·0	15·0	7·0
	23·0	6·0	7·5	7·4
	28·0	9·7	18·0	8·0
	17·0	3·0	7·5	3·0
	25·0	5·5	15·0	7·8
	15·0	6·0	5·0	0·7
	22·0	2·5	7·0	0·3
	19·0	4·0	14·5	7·5
	28·0	4·6	6·5	0·4
	27·0	3·1		
	26·0	4·8		
	17·0	2·5		
	25·0	3·6		
	17·0	2·2		
	17·5	5·2		
Total	334·5	70·7	96·0	42·1
Mean	22·30	4·71	10·67	4·68

Show that there is a significant difference between the two growth habits and that height contributes significantly to the discrimination of the two groups.

Show also that length minus height might be used as an index of growth habit.

72 A series of N observations classified in a 2×2 table gave the following numbers in each of the sub-groups:

a	b	$a+b$
c	d	$c+d$
$a+c$	$b+d$	N

Using a dummy variate to denote the grouping in either direction show that the correlation coefficient between the groupings in the two directions is

$$\frac{ad-bc}{\sqrt{[(a+b)(c+d)(a+c)(b+d)]}}$$

Since for N large and $P=O$, an estimated correlation coefficient tends to be normally distributed about zero mean with variance $1/N$, this implies that

$$\frac{N(ad-bc)^2}{(a+b)(c+d)(a+c)(b+d)}$$

is distributed as a $\chi^2_{(1)}$.

This example shows that the use of dummy variates for both dependent and independent variables leads to the χ^2 test provided sufficient observations are taken. It is, in consequence, a valid procedure provided sufficient observations are taken.

8

TRANSFORMATIONS AND NON-NORMAL DISTRIBUTIONS

8.1 *Reasons for transformations*—Application of many of the tests described in the previous sections depends upon one or more assumptions made concerning the nature of the observations. The main assumptions which are usually made may be listed as follows:

1 the residuals in the analysis are assumed to be normally distributed

2 the residual variances in groups to be compared are assumed to be equal

3 effects are supposed to manifest themselves by a constant increment to the observations so that the difference between group means indicates such effects

4 the means of each group are assumed to be indicative of the true mean.

These various assumptions are closely interwoven and the negation of one or two may imply the negation of another. As an illustration of the type of data for which these assumptions may not be true consider the comparison of two sets of large bacterial counts. Here the following difficulties are present:

i A series of counts is not usually normally distributed unless conditions are uniform. Very high counts often occur more frequently than in the normal distribution. For example, a series of counts may easily be 100, 1,000, 200, 500, 300, 50, 2,000, 400. These are obviously not normally distributed. If, however, the logarithms of these numbers are considered: 2·0, 3·0, 2·3, 2·7, 2·5, 1·7, 3·3, 2·6, these values are more nearly normally distributed.

ii The accuracy of the counts will usually depend upon the size of the count. Thus a group with a mean count of 1,000 may have a standard deviation of, say, 200. A group with a mean count of 100 will usually have a correspondingly smaller standard deviation of, say, 20. The fact that the standard deviation is proportional to the mean will tend to make unequal the residual variances in groups to be compared. If, however, the logarithms of the counts are used a small change in the logarithm causes a proportional change in the count, so that the standard deviation of the logarithms of the counts would be independent of the mean of the group. The logarithms of the counts might therefore be compared.

iii A method of treatment or a difference between groups will usually manifest itself by a proportional change in bacterial counts. Thus a treatment may tend to reduce counts by one half. The effect here cannot be represented by a constant increment. However, it has the effect of reducing the logarithms of the counts by log 2 so that the logarithms of the counts might be used.

iv The tendency for very high counts to occur makes the arithmetic mean an unsatisfactory index of the level of the whole group. For example, the arithmetic mean of the set of counts given in *ii* is 570 and is determined largely by the high count of 2,000. If the mean logarithm is used, this takes a value 2·51 corresponding to a count of 320. This is a better index of the group as a whole. To emphasize this point further suppose that the count of 2,000 had been 4,000 (which is quite possible) then the arithmetic mean would have been 820—a highly distorted value—but the corresponding figure of 350 from the transformed count would not have been so greatly influenced by this single high count.

It can be seen that each of these difficulties may be overcome by using the logarithms of the counts instead of the counts themselves. The counts are then said to be transformed using the logarithmic transformation.

The use of alternative measures in this manner is not restricted to logarithms, and various other measures will be suggested in the following sections to ensure that the main assumptions are justified. Unfortunately it is not always possible to satisfy all the assumptions at the same time and sometimes it is necessary to sacrifice some of them to achieve others. Usually it is more important to ensure that conditions *i* and *ii* are satisfied since the t test and variance-ratio test in the analysis of variance depend upon these two assumptions. However, sometimes the other two assumptions are of greater importance and these will also be considered in the following sections.

8.2 *Transformations to equalize variances*—Probably the most common type of transformation is that used to equalize variances or to make the variation in any group of observations independent of the mean. Usually this form of transformation also serves to make the distribution more nearly normal so that the t and variance-ratio tests may generally be applied without any difficulty. The different types of transformations and their effects will be considered in turn.

I Logarithmic transformation

As indicated above this is applied when the standard deviation is proportional to the mean. It is particularly useful for index numbers and quantities which tend to change proportionally.

The following data of tick counts on sheep (unpublished data of W. MOORE) demonstrate the use of the logarithmic transformation very well. The number of ticks was observed weekly on a group of 19 sheep previously sprayed with D.D.T. The mean and standard deviations of the tick counts were as follows:

Week	*First*	*Second*	*Third*	*Fourth*	*Fifth*
Mean	0·11	1·05	0·78	1·84	2·12
Standard deviation	0·31	1·28	0·81	1·64	1·32
Week	*Sixth*	*Seventh*	*Eighth*	*Ninth*	*Tenth*
Mean	2·53	2·44	4·32	3·63	7·00
Standard deviation	2·78	1·82	2·60	3·59	5·63

These results are plotted in *Figure 37*. It is seen that the standard deviation increases steadily with the mean showing that a logarithmic transformation is required. However, there were several occasions on which zero counts were observed so that the logarithms of these values could not be taken. To overcome this difficulty the logarithm of the count plus one was used. This is quite a common adjustment when low numbers are observed and it has almost the same effect as the straight logarithmic transformation. For the transformed counts the means and standard deviations were then:

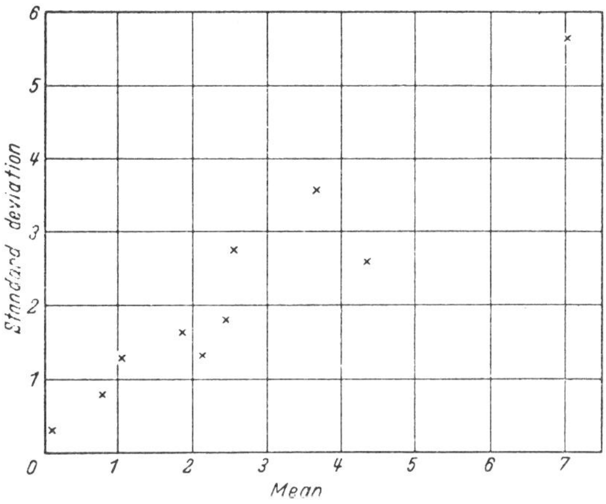

Figure 37. Plot of mean weekly tick counts against their standard deviations

Week	*First*	*Second*	*Third*	*Fourth*	*Fifth*
Mean	0·030	0·236	0·206	0·386	0·442
Standard deviation	0·09	0·26	0·20	0·25	0·24
Week	*Sixth*	*Seventh*	*Eighth*	*Ninth*	*Tenth*
Mean	0·430	0·485	0·651	0·567	0·807
Standard deviation	0·33	0·22	0·30	0·30	0·30

These figures do not present any systematic change in the standard deviation which, with the exception of the first week, is relatively constant at about 0·27. The figure for the first week, which results from two sheep with a count of 1 and seventeen with a count of 0, must necessarily remain low and this week should be isolated in further comparisons. The remaining weeks might now be analysed jointly.

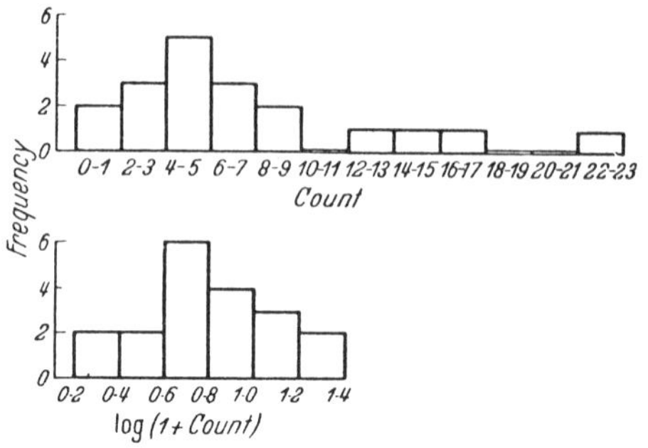

Figure 38. Histograms of untransformed and transformed sheep tick counts

As mentioned above, the logarithmic transformation tends to produce normality in the data at the same time as equalizing the variances. This is demonstrated in *Figure 38* which gives the histograms of the untransformed and transformed counts for the tenth week. The untransformed counts have a decided skewness indicated by observations straggling well above the centre of distribution. Although the histogram of the transformed counts is slightly irregular it is not nearly as skew as previously and the transformed values fall within a fairly narrow range. The deviation from normality for these transformed values is not significant.

II Square root transformation

It often happens that the standard deviation tends to increase with the mean but not proportionally. The most common alternative is that the variance increases proportionally with the mean. This usually occurs with colony counts* or with low counts of discrete numbers. As an example of this transformation, consider the following data of the numbers of deaths in litters reared on four different diets. The means and variances (with 21 d.f.) of the numbers were:

Diet	*1*	*2*	*3*	*4*
Mean	2·91	3·77	2·61	4·32
Variance	8·94	15·71	10·16	18·61

These are plotted in *Figure 39*. The increase in variance with the mean may be represented by a straight line of the type shown so that the square

* As shown in section 2A.9 these follow the Poisson distribution for which the variance is exactly equal to the mean.

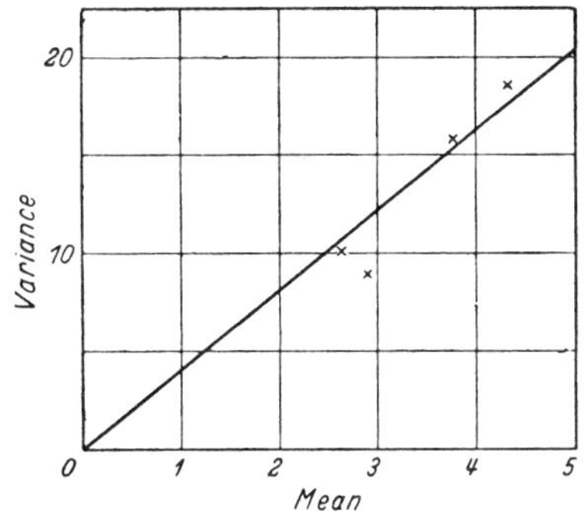

Figure 39. Plot of mean and variance of deaths on four diets

roots of the numbers of deaths might be used. If this is done the transformed means and variances become:

Diet	*1*	*2*	*3*	*4*
Mean	1·37	1·65	1·28	1·75
Variance	1·09	1·09	1·03	1·33

The variance is now virtually independent of the mean.

The difference between the means on these diets was not significant so that to demonstrate the form of distribution the individual figures might be combined. *Figure 40* indicates the form of the distribution of the untransformed results; it is seen that again the transformation reduces the skewness of the distribution and gives rise to a near normal form of distribution.

As with the logarithmic transformation when small numbers are involved it is better to add a constant before carrying out the transformation. F. J. ANSCOMBE has shown [*Biometrika* 35 (1948) 246] that the square root of the number plus 3/8 is the most suitable form of transformation.

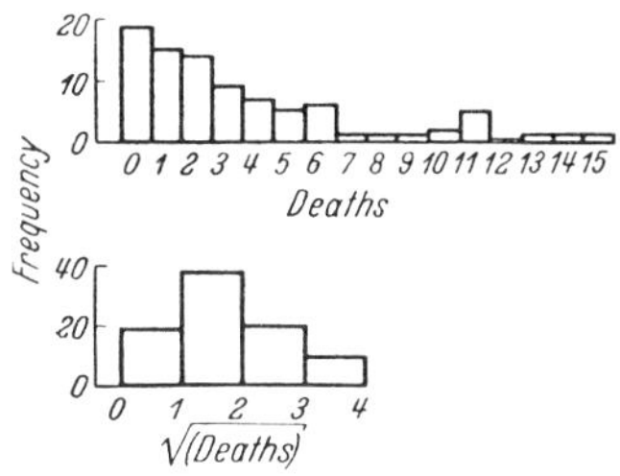

Figure 40. Histograms of untransformed and transformed numbers of deaths

As a further demonstration of the square root transformation, consider the following (unpublished) data of worm egg counts on sheep taken by G. C. HUNTER. Four worm egg counts were taken on each of 144 sheep, so that from each sheep a mean square with 3 degrees of freedom could be estimated. The following table groups the data according to the mean count from each sheep:

Grouping interval	*Mean count in interval*	*Estimated variance*	*D.f.*
0·00	0·00	0·00	36
0·25	0·25	0·25	45
0·50	0·50	0·40	57
0·75	0·75	1·19	51
1·00-1·25	1·10	1·28	39
1·50-1·75	1·61	1·10	60
2·00-2·75	2·22	2·47	57
3·00-4·75	3·85	4·35	39
5·00-	11·06	13·72	48

The means and variances of this table are plotted in *Figure 41*. It may be seen that the variance increases proportionally with the mean. The appropriate transformation is thus the square root. The 3/8 may or may not be added before applying the transformation. The following table gives the results obtained by both methods:

√(*Count*)		√(*Count*+3/8)		*D.f.*
Mean	*Estimated variance*	*Mean*	*Estimated variance*	
0·00	0·00	0·61	0·00	36
0·25	0·25	0·75	0·08	45
0·48	0·35	0·89	0·12	57
0·57	0·55	0·97	0·24	51
0·84	0·50	1·14	0·23	39
1·18	0·26	1·37	0·15	60
1·34	0·55	1·53	0·35	57
1·91	0·25	2·01	0·22	39
3·21	0·27	3·27	0·26	48

Both methods yield fairly stable variances, the correction of 3/8 giving a more constant variance for means of 1·00 and over. Again, for very low means the variance must necessarily remain low. The success of this transformation in stabilizing the variances indicates that the square root transformation should be used for making comparisons of worm egg counts on the same sheep. It does not, however, throw any light upon the appropriate method of making comparisons between different sheep since the above variances were estimated from counts taken on the same animals.

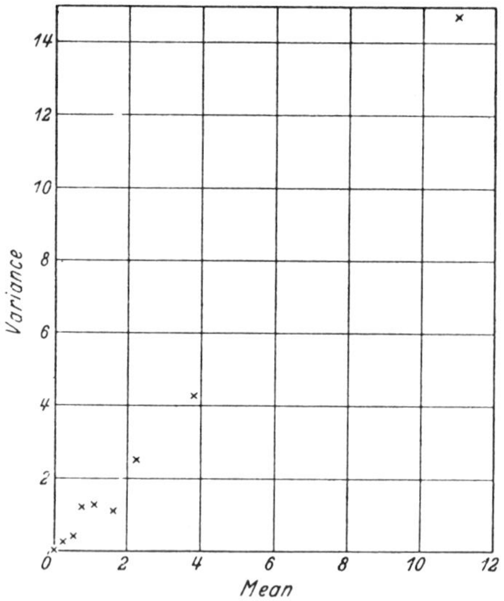

Figure 41. Plot of means and variances of worm egg counts

If it is desired to compare counts taken on different sheep, the manner in which the variance between sheep changes should be studied. Here this was done by dividing the animals into groups of six and estimating the means and variances of the total worm egg counts in each group. The means were then grouped according to size and the following values were obtained for the standard deviations in each group:

Grouping interval	*Mean count in interval*	*Estimated standard deviation*	*D.f.*
0-	1·3	1·0	5
2-	4·5	4·5	45
6-	8·0	10·2	35
10-	18·5	24·1	35

This table shows a proportional increase in the standard deviation with the mean. This indicates that, for comparing counts on different sheep, the logarithmic, and not the square root, transformation should be used. This demonstrates the need for care in ensuring that the estimated variances are applicable to the comparisons which are to be made.

III Sin$^{-1}\sqrt{p}$ transformation

The comparison of percentages should normally be carried out using the χ^2 test. However, if an extensive analysis is required this cannot be carried out very easily using χ^2. Thus, for example, if the percentages of diseased plants on different field plots were observed it may be desired to eliminate block differences before making any comparisons.

Since observed percentages tend to be normally distributed the possibility of carrying out an analysis of variance or t test using these percentages might be considered. However, such an analysis is complicated by the differing accuracies of the percentages. A low percentage can be determined with a greater accuracy than a medium percentage. For example, if out of 100 measurements only 1 has a given trait, we may be relatively certain that the true percentage lies between 0·1 and 5 per cent. However, if 50 were observed with the trait the corresponding limits would be 40 and 60 per cent. This does not mean that the standard deviation or variance increases with percentage, since if 99 out of 100 are observed with the trait the limits for the true percentage are again narrow. The accuracy here is the same as observing the percentage without the trait *i.e.* 1 out of 100.

As a consequence of this changing accuracy it is necessary to employ a transformation of the percentages to make the variance independent of the mean. This transformation is the $\sin^{-1}\sqrt{p}$ *i.e.* the angle of which the sine is the square root of the percentage. This quantity has been tabulated in radians in *Table IX* of the Appendix. A more extensive tabulation in degrees is to be found in FISHER, R. A. and YATES, F. *Statistical Tables for Biological, Agricultural and Medical Research.*

To demonstrate the effect of this transformation a series of (unpublished) observations on wild white clover taken by J. L. DAWSON and K. R. WILSON will be considered. Here, five estimates of the percentage of wild white clover were made on each of 60 plots, each estimate being made visually to the nearest 10 per cent. The means and variances of each set of estimates were calculated and are presented in the following table, grouped according to the mean percentage:

Grouping interval	*Mean percentage in interval*	*Estimated variance*	*D.f.*
0%-	2·7	4·0	12
10%-	12·0	12·0	8
20%-	25·0	18·6	40
30%-	34·3	18·9	108
40%-	44·7	23·4	72

These means and variances are plotted in *Figure 42*. It is seen that the variance increases with the mean observed percentage, but that for percentages between 25 and 50 it is relatively constant. This is characteristic of the manner in which the variance changes with the observed percentage. For percentages over 50 the variance decreases again to zero, slowly between 50 and 75 per cent but more rapidly between 75 and 100 per cent.

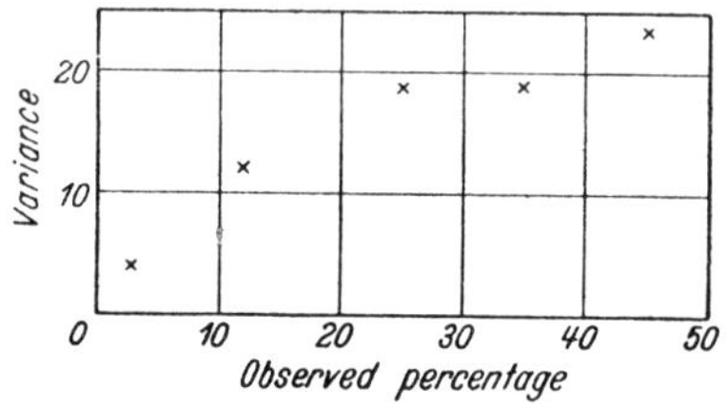

Figure 42. Plot of mean percentages and variances of wild white clover observations

These percentages were transformed by the $\sin^{-1}\sqrt{p}$ transformation and the variances for each set calculated as above. These transformed means and variances were then:

Mean	*Variance*	*D.f.*
0·07	0·028	12
0·31	0·042	8
0·50	0·037	40
0·62	0·025	108
0·74	0·030	72

The variances in this table are much more stable and they show no systematic trend with the mean. The transformation has largely succeeded in stabilizing the variance.

It must be noted that, as in the example of the last section, this approach only shows that the $\sin^{-1}\sqrt{p}$ transformation is appropriate for comparisons within the plots. To test whether the transformation is appropriate for comparing the percentages between the plots it is necessary to obtain estimates of the variance between the plots and to determine how these vary with the mean percentage. Here there were not sufficient plots to allow this to be done for the wild clover alone, but the overall results for the four main species were:

Mean percentage	*Estimated variance plot*
3·0	7·3
13·2	29·7
34·1	52·8
38·1	47·6

These results, plotted in *Figure 43*, show the same trend as those in *Figure 42* although the variation between plots is roughly double that within plots. The use of this transformation in analysing the variations of mean percentages per plot would thus seem justified.

Various points should be noted concerning the use of the $\sin^{-1}\sqrt{p}$ transformation:

1 While this transformation makes the variance independent of the observed percentage, it still depends upon the number of observations. Thus the transformation is most useful when the percentages are based upon equal numbers. Otherwise, the analysis of the transformed percentages should take into account the differing numbers upon which the percentages are based.

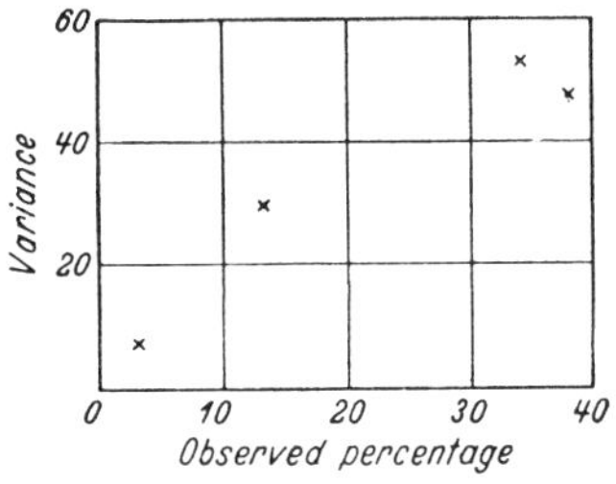

Figure 43. Plot of mean percentages and variances of observations on different species

2 For small percentages the variance is roughly proportional to the mean percentage so that the transformation is similar to the square root transformation. For larger percentages the variance falls off in proportion to the mean percentage.

3 Although this transformation makes the residual variance constant, it only achieves in part the other conditions. It will be seen later that other transformations have to be used for these purposes.

As a consequence of these various limitations, the $\sin^{-1}\sqrt{p}$ transformation is not applicable as widely as the logarithmic or square root transformations.

IV $\frac{1}{\beta}\sinh^{-1}\beta\sqrt{x}$ *transformation*

When the standard deviation increases roughly linearly, but not proportionally, with the mean, the logarithmic transformation may no longer be used. As indicated in section 8.2 *I* the logarithm of the measurement plus one *i.e.* $\log(x+1)$, may be employed instead, but there is an alternative transformation which may often be more useful. This alternative replaces the measurement x by the value of $\frac{1}{\beta}\sinh^{-1}\beta\sqrt{x}$, where β is the slope of the line giving the dependence of the standard deviation on the mean. Usually, however, for this rather complicated transformation to be worthwhile the form of dependence of the standard deviation on the mean has to be of a particular type:

1 For a mean less than one, the variance has to be roughly proportional to the mean *i.e.* the distribution behaves as in section 8.2 *II*.

2 For a mean greater than one, the standard deviation has to increase roughly linearly with the mean.

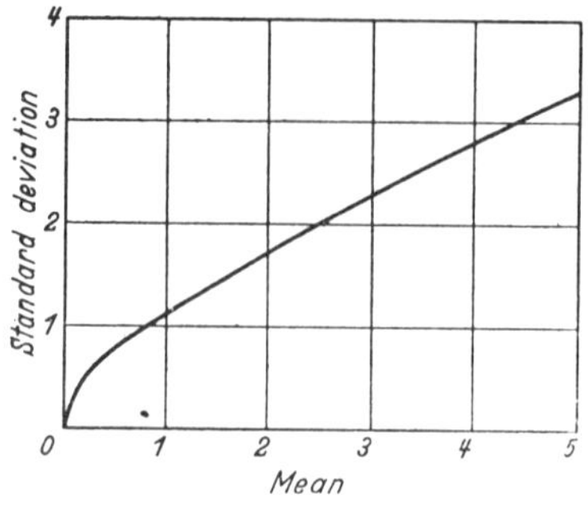

Figure 44. Type of dependence of standard deviation which requires $\frac{1}{\beta}\sinh^{-1}\beta\sqrt{x}$ *transformation*

Figure 44 gives an illustration of the type of dependence. The slope β of the line in this figure is 0·5.

This type of transformation is of most use where the observations are counts in which the objects counted tend to group together. If there is no grouping, the square root transformation is appropriate and this is equivalent to setting $\beta=0$. If grouping occurs, then the value of β acts as an index of the intensity of such grouping. Thus, in the example of section *I*, since the ticks occur in colonies the tick counts tend to group on some sheep. It might therefore be expected that the comparison of counts on different sheep would require a transformation of the type $\frac{1}{\beta}\sinh^{-1}\beta\sqrt{x}$. Closer inspection of *Figure 37* shows that it does seem to possess the same tendency as *Figure 44* for low counts. In consequence the $\frac{1}{\beta}\sinh^{-1}\beta\sqrt{x}$ transformation might be used with β equal to the slope of the line *i.e.* about 1·1.

Normally, unless many small observations are taken and unless there are indications of the above type that this transformation is required, it is easier and almost as good to employ either the transformation log $(1+x)$ or the transformation $\sinh^{-1}\sqrt{x}$ *i.e.* setting $\beta=1$. This latter transformation is tabulated in *Table X* of the Appendix. It may be usefully employed for quite a wide range of values of β.

If the transformation $\sinh^{-1}\sqrt{x}$ is used to transform the tick counts, the transformed means and standard deviations become:

Week	*First*	*Second*	*Third*	*Fourth*	*Fifth*
Mean	0·093	0·535	0·514	0·908	0·982
Standard deviation	0·28	0·60	0·51	0·49	0·51
Week	*Sixth*	*Seventh*	*Eighth*	*Ninth*	*Tenth*
Mean	0·931	1·118	1·297	1·212	1·591
Standard deviation	0·65	0·39	0·54	0·52	0·38

These values behave in the same manner as those derived using the logarithmic transformation. Here the difference between the standard deviation for the first week and those for the remaining weeks is not as great as previously, but they still differ by a factor of nearly two. Evidently this transformation does not offer any advantage over the logarithmic transformation in this instance.

It should, however, be noted that the second lowest mean count (in the third week) no longer has the second lowest standard deviation. The $\sinh^{-1}\sqrt{x}$ transformation has apparently removed this effect and if there had been several mean counts between 0·1 and 0·8 it might have shown an appreciable advantage over the logarithmic transformation. Such a state of affairs was, in fact, observed on a second group of twenty sheep which had been dipped. In the first week after dipping the means and standard deviations of the untransformed and transformed counts were:

	Untransformed	*Transformed*	
		Using $\log(1+x)$	*Using* $\sinh^{-1}\sqrt{x}$
Mean	0·45	0·114	0·300
Standard deviation	0·82	0·18	0·48

Here the $\sinh^{-1}\sqrt{x}$ transformation succeeds in making the variance equal to the others while the $\log(1+x)$ fails.

V Reciprocal transformation

It has been shown that if the variance tends to increase proportionally with the mean the square root transformation should be employed. If, alternatively, the variance tends to increase as the square of the mean *i.e.* the standard deviation tends to increase proportionally with the mean, the logarithmic transformation is required. If, however, the variance tends to increase still more rapidly with the mean *i.e.* if the standard deviation increases more than proportionally with the mean, a further transformation may be required. The most common transformation here is the reciprocal transformation in which each measurement is replaced by its reciprocal. This transformation is appropriate when the variance increases as the fourth power of the mean—a very great rate of increase.

It is very seldom that the variance increases as rapidly as this, but such a state of affairs might be expected to occur when size implies a highly variable structure. This may occur in some economic measurements where, for example, an increase in capacity to produce is followed by 'overproduction' and a consequent increase in variability, but the number of such practical examples requiring this transformation is fairly small.

The effect of the transformation is to reverse the order of measurements. Thus very large values become very small and very small values become very large. Also it cannot be applied where zero values are observed. Here, however, the reciprocal of the value plus one *i.e.* $1/(x+1)$, may often be used.

8.3 *Transformations to achieve normality*—Transformations may often be employed to ensure that the distribution of observations is fairly normal. Fortunately, if the distribution does not deviate greatly from normality, the effect may often be neglected in subsequent analysis. Thus the mean of a number of observations tends to be normally distributed whether the individual observations are normally distributed or not, and the variance-ratio test for testing the difference between group means is valid for large groups for all the distributions commonly met in practice. As a consequence, the attainment of normality is usually a second consideration after the equalization of variances and provided the observations do not show very marked deviations from normality this consideration need not cause undue concern. However, some care has to be taken where comparisons are being made with small numbers of observations.

Fortunately, if the distribution does not deviate greatly from normality, transformations to equalize variances are also those which may be used to achieve normality. As a result, the same transformation often succeeds in both objectives simultaneously. The common normalizing transformations will be considered in turn.

I Logarithmic, square root and reciprocal transformations

The logarithmic, square root and reciprocal transformations may be used to correct skewness in any set of positive observations. Most usually these transformations are applied to counts; but their application is not restricted to counts. They cannot, however, be applied to negative observations*.

The square root transformation is the least drastic of the three transformations and is applied to moderately skew distributions. Its suitability may be rapidly and roughly gauged by calculating the numbers of observations falling in the groups: 0-, 1-, 4-, 9-, 16- *etc.* These frequencies should be approximately symmetrical. Alternatively, it is often more useful to consider the manner in which the centre *i.e.* the median and upper and lower 10 per cents of the distribution will transform.

Figure 45a gives a distribution to which the square root transformation might be applied. The centre of the distribution falls between 4 and 5, say 4·5, about 12 per cent of the observations take a value 2 or less, and about 12 per cent take a value 8 or more. The square roots of 4·5, 2 and 8 are 2·1, 1·4 and 2·8 respectively, and these are equally spaced, indicating the suitability of the square root transformation.

The logarithmic transformation is more drastic and should be applied to

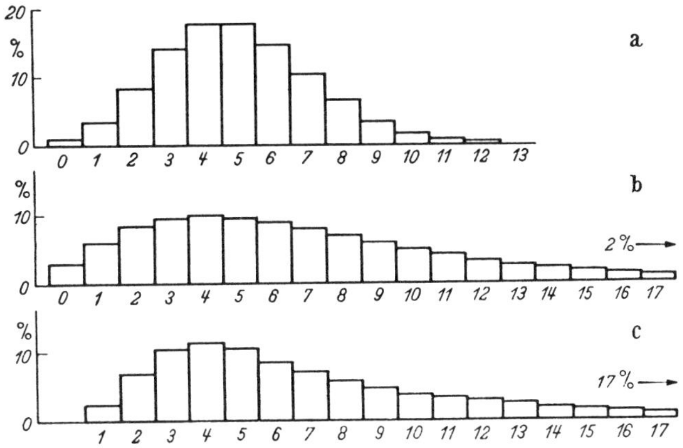

Figure 45. Types of skew distributions

fairly skew distributions. Its suitability may be gauged by calculating the numbers* of observations falling into the groups 0-, 1-, 2-, 4-, 8-, 16- *etc.* These frequencies should be approximately symmetrical. Again, as above, the use of central and extreme values acts as a good indicator of the suitability of this transformation.

Figure 45b illustrates the type of distribution to which the logarithmic transformation should be applied. Here the centre of the distribution is at about 5·0, 9 per cent of observations take a value 1 or less and 10 per cent take a value 14 or more. The logarithms of one plus these values are 0·78, 0·30 and 1·18 which are approximately equally spaced.

The reciprocal transformation is the most drastic of these transformations. It requires a very skew distribution with no zero observations to justify its use and even then it is always advisable to test that the logarithmic transformation is not suitable before applying the reciprocal transformation. As for the other two transformations, the central and extreme observations may be used to indicate its suitability.

Figure 45c gives a distribution for which the reciprocal transformation might be used. For this distribution, the centre lies at about 5, 9 per cent of observations are 2 or under and 9 per cent are 25 or over. The reciprocals of these values are 0·2, 0·5 and 0·04. These are not equally spaced and here the reciprocal of the value plus one would be more suitable. Alternatively, the logarithmic transformation might suffice. The logarithms of 5, 2 and 25 are 0·7, 0·3 and 1·4 respectively, and these are approximately equally spaced.

As an example of the choice of transformations consider the following

* For the logarithmic transformation this difficulty may be overcome by using $\sinh^{-1}x$ instead of $\log(x+1)$.

distribution of numbers of 'non-resident' species of birds observed on certain days in September over several years:

Number of non-resident species	1	2	3	4	5	6	7	8
Frequency	6	6	15	26	22	20	13	14
Number of non-resident species	9	10	11	12	13	14	15	16
Frequency	9	7	2	7	4	3	0	1
Number of non-resident species	17	18	19	20	21	22	*Total*	
Frequency	2	1	0	1	0	1	160	

This distribution resembles that of *Figure* 45b. Its centre lies at about 5·5, 12 observations are 2 or less and 13 observations are 13 or more. If the logarithm of the number of species plus one is taken (since there is a possibility of observing no species) these three values become 0·81, 0·48 and 1·15 respectively. These are almost equally spaced so that this transformation seems to be suitable.

In practice there are seldom sufficient observations to enable the distribution to be represented accurately by a histogram. Consequently the form of the distribution has to be gauged from a few observations. This can be done by a consideration of the highest and lowest observations. For example, the following (unpublished) series of cone counts was made by W. NEIL on ten pine trees: 0, 28, 67, 81, 120, 137, 337, 386, 564, 569. The centre of these values is about 128 and the counts 28 and 564 may be taken as representing typical low and high counts. The square roots of the values 28, 128 and 564 are approximately 5, 11 and 24 so that this transformation is not sufficiently drastic. The logarithms of these values are 1·45, 2·11 and 2·75. These are very nearly equally spaced so that the logarithmic transformation appears to be suitable. Since a zero count occurs it is necessary to employ the logarithms of the counts plus one.

II Transformation of ranks to normal scores

Often, although it is impossible to express a set of observations as measurements, the observations may be arranged in order. Thus, for example, beauty cannot be measured but a series of observations may be arranged in order of beauty. The position of any observation in such an order is said to be its rank. Thus the first has rank 1, the second rank 2 and so on. Exact tests exist using such ranked observations* but transformed ranks may often be utilized in carrying out an analysis. A suitable transformation for this purpose appears in FISHER, R. A. and YATES, F.

* KENDALL, M. G. *Rank Correlation Methods* London, 1948

Statistical Tables for Biological, Agricultural and Medical Research London, 1943. The average value of the *r*th largest of *n* observations taken from a normal distribution has been tabulated. Thus, in the analysis of a set of *n* ranks, each rank is replaced by its corresponding normal score. As an example consider the following set of observations of the relative abilities in English and Arithmetic of 10 boys:

Boy	*1*	*2*	*3*	*4*	*5*	*6*	*7*	*8*	*9*	*10*
Position in Arithmetic	8	10	5	4	3	2	6	1	9	7
Position in English	7	9	8	1	4	2	5	3	10	6

To test now whether the positions in English and Arithmetic are related it is necessary to calculate the correlation coefficient between the two sets of observations. However, in order to make the test of the correlation valid it is necessary to replace these by their corresponding normal scores: 1 is replaced by 1·54, 2 by 1·00, 3 by 0·66, 4 by 0·38, 5 by 0·12, 6 by −0·12, 7 by −0·38, 8 by −0·66, 9 by −1·00, and 10 by −1·54. The table then becomes:

Boy	*1*	*2*	*3*	*4*	*5*	*6*	*7*	*8*	*9*	*10*
Score in Arithmetic	−0·66	−1·54	0·12	0·38	0·66	1·00	−0·12	1·54	−1·00	−0·38
Score in English	−0·38	−1·00	−0·66	1·54	0·38	1·00	0·12	0·66	−1·54	−0·12

The correlation coefficient testing the association between score in Arithmetic and score in English is 0·773 and this may be tested using the variance ratio

$$\frac{8\,(0{\cdot}773)^2}{1-(0{\cdot}773)^2}=11{\cdot}9\,,$$

with 1 and 8 degrees of freedom. Referring to *Table IV* it is seen that as high a value would occur by chance less than once in a hundred times. The association between the two ranks is evidently significant.

8.4 *Numerical presentation of transformed data*—When an analysis has been carried out using transformed data there is usually some difficulty concerning the presentation of results. Thus we may obtain means, differences between means and standard errors for the transformed data and then require to transform back to original scale to interpret the results and possibly to compare them with untransformed values. This presents several problems which will be considered in turn.

First, the means have to be transformed back to the original scale. For example, the antilogs of the mean transformed tick counts from sheep of the data of section 8.2 may be calculated and 1 subtracted from each to give:

Week	*First*	*Second*	*Third*	*Fourth*	*Fifth*
Count	0·07	0·72	0·61	1·43	1·77
Week	*Sixth*	*Seventh*	*Eighth*	*Ninth*	*Tenth*
Count	1·69	2·05	3·48	2·69	5·41

In the same manner, mean square roots would have to be squared, mean reciprocals inverted and so on. The values derived in this manner, which we shall call derived means, will usually be less than the means calculated by the direct method since they will not be so greatly affected by extreme observations.

Secondly, the comparison of any pair of means should be carried out on the transformed data and the level of significance determined. This should be used in general as indicating the significance of the comparison of the derived means. Confidence limits may be set for any of the transformed means and these, when transformed back to the original scale, give the corresponding limits for the derived means. Fortunately, standard errors derived using the logarithmic transformation are capable of special interpretation. Since the logarithm reflects proportional differences the antilogarithm of the standard error of a mean indicates the proportional variation in the derived mean. This is not true, however, when the logarithm of one plus the measurement is used.

Thirdly, in order to make the derived means comparable with the ordinary values calculated by straight averaging, but not liable to be greatly distorted by one or two extreme values, it is necessary to make certain adjustments according to the form of the transformation:

1 for the logarithmic transformation (with or without the 1), 1·15 times the variance of the observations should be added to the transformed mean before transforming back

2 for the square root transformation (with or without the $\frac{3}{8}$) the variance of the transformed observations should be added to the derived means.

Corresponding, but more complicated, adjustments exist for the $\sin^{-1}\sqrt{p}$ and $\sinh^{-1}\sqrt{x}$ transformations*.

If such an adjustment is applied to the tick counts from sheep, 1·15 × (the pooled estimate of variance for the transformed counts in the second to

*If $\bar{t}$ is the transformed mean and s^2 the variance of the transformed observations, $\cos 2\bar{t}.(1-e^{-2s^2})/2$ and $\cosh 2\bar{t}.(e^{2s^2}-1)/2$ have to be added to the derived means for these transformations.

tenth weeks)=0·084 is added to each mean before transforming back. This gives a revised table of means for these weeks:

Week	*First*	*Second*	*Third*	*Fourth*	*Fifth*
Count	—	1·09	0·95	1·95	2·36
Week	*Sixth*	*Seventh*	*Eighth*	*Ninth*	*Tenth*
Count	2·27	2·71	4·43	3·48	6·78

These are in quite good agreement with the values obtained by direct averaging. However, for these values it is possible to test which means differ significantly. For the first week where the variance was different from the rest, a separate adjustment has to be made. This gives a value 0·10 for the mean count as previously.

As a second example, consider the data on deaths used to illustrate the square root transformation. The squared means of the transformed values are 1·877, 2·722, 1·638 and 3·062 respectively and to each of these must be added the pooled estimate of variance, 1·135. The table of derived means is then:

Diet	1	2	3	4
Mean	3·01	3·86	2·77	4·20

These values do not differ very greatly from the direct averages, but the significance of the differences between these means can be tested using the transformed values. The differences between these means and those calculated directly might be considered as adjustments to eliminate the effects of extreme observations.

8.5 *Graphical presentation of transformed data*—The presentation of transformed data in scatter diagrams may be carried out by transforming and plotting the data. Often however special graph paper is available to avoid the necessity of carrying out the transformation and to allow the values of the observations to be read directly from the graph. This leads to greater ease in transferring the data to the graph and reading results from the graph.

Figure 46 shows a plot of the cost of living index and average daily earnings in industry in Chile between 1937 and 1948. This plot shows that the proportional increase in both over this period tended to be constant but the increasing distance between the measurements shows that earnings are increasing proportionately more than the cost of living. This graph, which employs a logarithmic scale, is relatively easy to interpret, although there is a tenfold range in the values plotted.

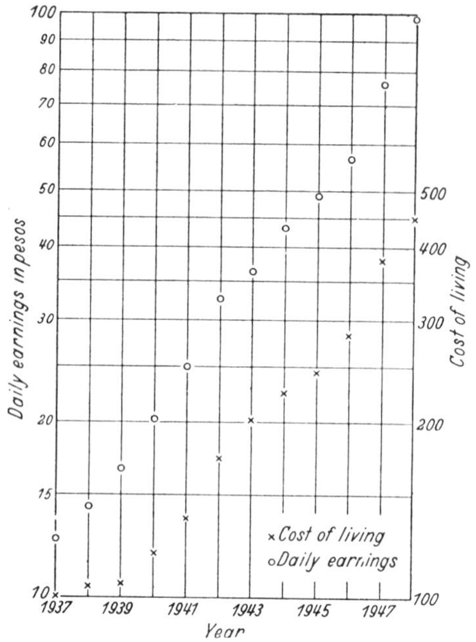

Figure 46. Cost of living and average daily earnings in Chile from 1937 to 1948

In a similar manner, suitable scales may be chosen for the square root or reciprocal transformations. The logarithmic scale is, however, the most widely applicable and graph paper may be obtained to represent proportional changes of up to a million on a logarithmic scale. Thus, for example, *Figure 47* shows the change in the cost of living index in France between 1937 and 1948 on a scale suitable for a 100-fold increase. This graph shows quite clearly that the rate of increase in the cost of living was increasing between 1938 and 1945, but that after this date the rate of increase was constant.

Use of scales of this type is not restricted to one variable only and the scales in both directions may be altered to effect transformations. Again, use of a logarithmic scale in both directions is most common and this has many applications. In particular, if the variances and means of a set of data are plotted on graph paper with logarithmic scales, the most suitable form of transformation can be determined from the slope of the resulting line. A slope of 1 indicates the suitability of the square root transformation while a slope of 2 indicates that the logarithmic transformation should be applied.

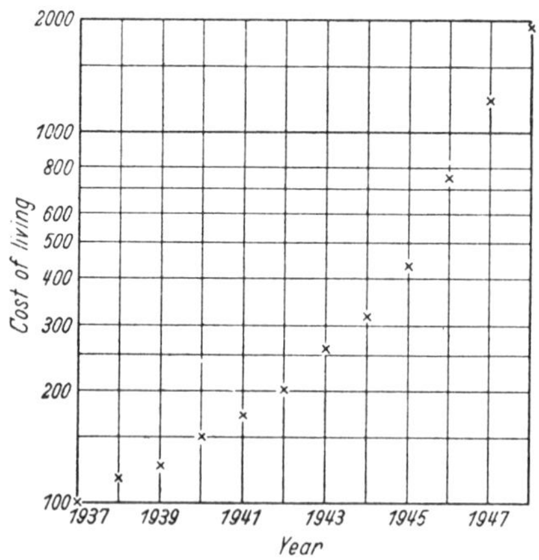

Figure 47. Cost of living index in France from 1937 to 1948

Figure 48 gives a plot of the variances and means of the worm egg counts of section 8.2 on logarithmic graph paper. The slope of this line is 1 showing that the square root transformation is required.

The variances and means of the tick counts from sheep of section 8.2 are likewise plotted in *Figure 49*. Here the slope of the line is 2 showing that the logarithmic transformation is necessary. It must, however, be noted that the slope of the line joining the first two points in this figure is more nearly 1 than 2. This indicates that the transformation $\frac{1}{\beta}\sinh^{-1}\beta\sqrt{x}$ might alternatively be applied.

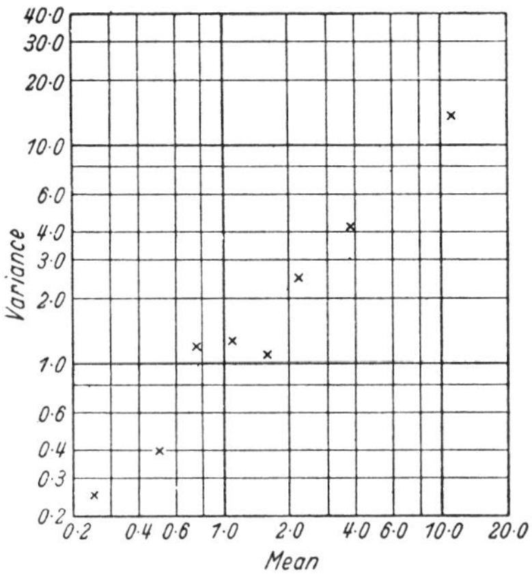

Figure 48. Plot of means and variances of worm egg counts

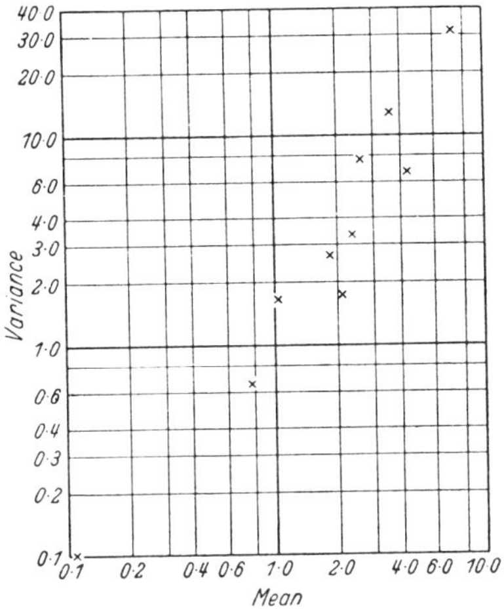

Figure 49. Plot of means and variances of tick counts from sheep

SUMMARY OF PP 163 TO 181

Reasons for the application of transformations have been given and the various transformations appropriate for particular purposes have been demonstrated.

Five transformations have been given to make variances independent of the means: the logarithmic, square root, $\sin^{-1}\sqrt{p}$, $\frac{1}{\beta}\sinh^{-1}\beta\sqrt{x}$ and the reciprocal, and the suitability of each demonstrated.

The applications of the logarithmic, square root and reciprocal transformations in achieving normality have been illustrated and the use of normal scores in analysing ranks has also been indicated.

Finally, the methods of presenting transformed data numerically and graphically have been shown.

EXAMPLES

73 The numbers of non-resident species of birds were observed over a series of years. The following figures give the means and variances of these numbers according to the weeks in which they were observed:

	March 3	March 4	April 1	April 2	April 3	April 4	May 1	May 2	May 3	May 4	June 1	June 2
Mean	5·6	6·0	7·5	8·9	10·8	5·9	11·9	15·7	9·7	6·3	2·3	1·5
Variance	6·8	14·9	26·4	18·2	18·5	11·9	60·9	66·4	32·1	8·3	8·9	6·1

	August 2	August 3	August 4	September 1	September 2	September 3	September 4	October 1	October 2	October 3
Mean	2·7	3·6	5·0	6·5	6·4	7·1	6·6	10·5	9·4	8·9
Variance	2·6	5·7	5·2	5·9	8·0	25·6	16·0	31·7	10·6	8·5

Show that these figures suggest use of the logarithmic transformation.

74 The following table gives the hourly lemon sole catches of two research vessels—the *Explorer* and the *Scotia*—fishing in the same area on three successive days:

Day	*Explorer* 1	*Explorer* 2	*Explorer* 3	*Scotia* 1	*Scotia* 2	*Scotia* 3
	57	41	28	6	3	5
	31	14	35	4	3	13
	20	38	11	5	1	9
	37	17	18	19	6	6
	22	21	18	8	2	12
	31	10	13	12	5	5
Mean	33·0	23·5	20·5	9·0	3·3	8·3
Variance	178·0	167·4	85·0	32·0	3·4	12·6

Show that the logarithmic transformation should be used in analysing this data. Using the logarithm of each catch plus one calculate the following transformed means and variances:

	Explorer 1	*Explorer* 2	*Explorer* 3	*Scotia* 1	*Scotia* 2	*Scotia* 3
Mean	1·51	1·34	1·30	0·95	0·60	0·94
Variance	0·0252	0·0526	0·0330	0·0499	0·0399	0·0271

Test the differences between the catches by the two vessels and on different days using the analysis of variance:

	D.f.	*S.s.*	*M.s.*	*V.r.*
Vessels	1	2·7280	2·7280	72·0
Days	2	0·4027	0·2014	5·31
Vessels × days	2	0·2168	0·1084	2·86
Groups	5	3·3475	—	—
Residual	30	1·1383	0·0379	
Total	35	4·4858		

Hence construct the following tables of adjusted means:

Vessel	*Explorer*	*Scotia*
Mean	26·62	6·78

Day	1	2	3
Mean	22·44	11·94	17.38

Here the adjustments to the means are made using the variances calculated when only the effect to be presented is eliminated. Thus the estimate of variance used in presenting the vessel means is $(4{\cdot}4858-2{\cdot}7280)/34=0{\cdot}0517$ and the estimate used in presenting the daily means is $(4{\cdot}4858-0{\cdot}4027)/33=0{\cdot}1237$. The residual mean square would be the appropriate estimate of variance where the individual group means were to be constructed and compared.

75 Transform the data of example *3* using the $\sin^{-1}\sqrt{p}$ transformation and test the difference between the percentages of seeds germinating in the two mixtures by a t test ($t_{(18)} = 3{\cdot}44$).

76 Re-analyse the data of example *35* using the $\sin^{-1}\sqrt{p}$ transformation and show that the conclusions reached in the previous analysis are unaltered.

77 The following table gives the regeneration counts of Scots pines in four blocks of six plots. Three plots in each block were poor in heather status and the other three were rich.

Block	*1*		*2*		*3*		*4*	
Heather status	*Poor*	*Rich*	*Poor*	*Rich*	*Poor*	*Rich*	*Poor*	*Rich*
	24	17	3	0	12	1	2	5
	2	4	0	0	9	3	8	2
	14	0	2	2	17	0	0	4

Obtain an estimate of the standard deviation for each block (based upon 4 degrees of freedom) and derive the following table:

Block	*1*	*2*	*3*	*4*
Mean	10·2	1·2	7·0	3·5
Standard deviation	10·7	1·5	7·6	3·1

Hence show that the $\log(1+x)$ transformation is required.

Using this transformation construct the following analysis of variance:

	D.f.	*S.s.*	*M.s.*	*V.r.*
Heather status	1	0·55	0·55	3·44
Blocks	3	1·11	0·37	2·31
Residual	19	3·10	0·16	
Total	23	4·76		

The difference between the plots with rich and poor heather status is suggestive but not significant.

78 Plot the figures shown in the following table on a logarithmic scale and comment upon the relative trends:

Numbers of Divorces by Country

Year	*1937*	*1939*	*1941*	*1943*	*1945*	*1947*
England and Wales	5,045	7,019	6,318	10,724	18,982	52,249
Scotland	637	869	760	1,301	2,205	2,499
Northern Ireland	12	31	67	119	174	196

79 The following table gives the means and variances of colony counts taken from three soil samples [data of JONES, P. C. T. and MOLLISON, J. E. *J. gen. Microbiology* 2 (1948) 54]. Show that these suggest that the square root transformation should be applied:

Mean	2·68	5·94	2·79	2·95
Variance	2·55	4·43	2·98	2·16

80 The distribution of private incomes in 1946 was:

Income	*No. of incomes (thousand)*	*Income*	*No. of incomes (thousand)*
Under £250	?	£1,000-2,000	495
£250-500	6,600	£2,000-10,000	157
£500-1,000	1,740	£10,000 *and over*	8

Show that these figures suggest the use of the logarithmic transformation in any analysis involving private income. Check this inference using the data of example *10*.

81 Verify that the logarithmic transformation is suitable for the data of *Figure 13* and example *15*.

82 The following table gives the male death rates per 1,000 population in the United Kingdom in 1938 and 1946. Plot these figures on a logarithmic scale:

Age group	0-4	5-9	10-14	15-19	20-24	25-34
1938	18·0	1·9	1·3	2·0	2·7	2·9
1946	14·0	1·0	0·9	1·6	2·6	2·3

Age group	35-44	45-54	55-64	65-74	75-84	*Over* 85
1938	4·7	10·4	23·2	54·1	130·5	261·9
1946	3·7	9·2	22·5	50·3	118·0	238·0

83 The ranking in intelligence of the 10 boys tested in section 8.3 was 6, 9, 8, 1, 7, 2, 5, 3, 10, 4; show that the partial correlation between ability in English and ability in Arithmetic is 0·59 when the effect of intelligence has been removed. Show also that scores in English and Arithmetic can be used to account for 90 per cent of the variation in intelligence.

EXTENDED DEVELOPMENT

8A.6 *Additive transformations*—Sometimes the consideration that the measurement employed should be capable of simple interpretation and comparison is more important than any other consideration. Additive transformations might be used to achieve these objectives. This simplicity in interpretation is, however, often attended by complexity in analysis and such transformations are, in general, of limited value. Three main transformations to achieve additivity might be noted.

I Logarithmic transformation

This transformation is important whenever proportional changes occur in the measurements to be considered. For example, a change in an index number from 50 to 100 is comparable with a change from 100 to 200, and so on. Thus, if an index number varies in magnitude over a wide range, in order to determine comparable changes it is necessary to use the logarithm of the measurement as in *Figures 46* and *47*. The same use of logarithms is required in many economic and biological measurements taken over a wide area or a very long period of time.

II Probit transformation

This transformation is used to make changes in a series of percentages comparable. For example, if a drug reduces the percentage mortality from a disease from 99 to 98 per cent, this change is not comparable with a change from 51 to 50 per cent. While the former change doubles the number of survivals, the latter is, in fact, less spectacular and less important. To overcome this difficulty each percentage has to be transformed into its corresponding normal deviate. This might be done using *Table I*. This transformation is called the probit transformation.

The probit transformation is most useful where changes in percentage mortality or any other measured percentages are being studied in relation to changes in other factors such as dose. Consequently, the fitting of regression lines to data of this type is of some importance. The method of fitting cannot, however, be discussed here and reference should be made to specialized works on this subject for exact details*.

Here, in particular, the graphical presentation of results is important and special percentage or probability paper is available for this purpose. *Figure* 50 demonstrates the application of this to two sets of experimental results. *Figure* 50 shows the percentage mortality of mice classified according to the weight increases prior to infection. It is apparent that the low weight increases are associated with a high percentage mortality. *Figure* 50b shows the percentage mortality in two groups of sheep vaccinated with two different sera, toxoid and anaculture, and classified according to exposure to risk. One of these vaccinations, anaculture, is consistently better than the other and to about the same degree.

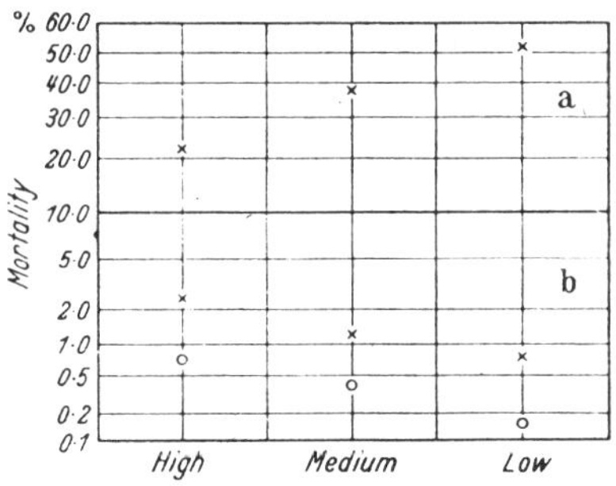

Figure 50. *Examples of use of probability paper*

III $-\text{Log}(1-p)$ *transformation*

This transformation is again suitable for percentages but for a different type of problem. Suppose in a series of observations on groups of individuals the presence or absence of a particular trait within each group is observed, but not the proportion having the trait. The percentage p of groups with the trait may then be used to indicate the percentage of individuals possessing the trait. This is done by using $-\log(1-p)$ instead of p. For example, if batches of articles are examined and the proportion

* FINNEY, D. J. *Probit Analysis* Cambridge, 1947, for example.

p containing defective articles estimated, then the proportion of defective articles is proportional to $-\log(1-p)$.

In particular, if two or more traits are observed simultaneously, then the values of $-\log(1-p)$ indicate the relative density of each trait and the values of $-\log(1-p)$ may be added to indicate the densities of the combined traits.

To demonstrate this transformation consider the following (unpublished) data collected by C. H. GIMINGHAM. The presence or absence of different species of plants was observed in a series of quadrats thrown at random. This was done on a series of plots subject to different grazing conditions. The following table gives a portion of the results:

Proportion p of Quadrats containing Species

Plot \ Species	*Carex*	*Luzula*	*Deschampsia*	*Agrostis*	*Vaccinium*
Ungrazed	0·650	0·275	0·400	0·075	0·125
Grazed all year	0·825	0·350	0·725	0·175	0·225

In order to gain some idea of the relative densities on the grazed and ungrazed plots, minus the logarithms of $(1-p)$ must be used. This gives:

Relative Densities of Species

Plot \ Species	*Carex*	*Luzula*	*Deschampsia*	*Agrostis*	*Vaccinium*
Ungrazed	0·456	0·140	0·222	0·034	0·058
Grazed all year	0·757	0·184	0·561	0·084	0·111
Grazed/Ungrazed	1·66	1·31	2·53	2·47	1·91

These figures show clearly the preponderance of *Carex* on both the grazed and ungrazed plots. They also show a greater density in each species on the grazed plots. Some or all of these species might now be combined in making comparisons. For example, the following table might be constructed:

Plot \ Species	*Carex*	*Other species*	*All species*
Ungrazed	0·456	0·454	0·910
Grazed all year	0·757	0·940	1·697
Grazed/Ungrazed	1·66	2·07	1·86

This shows more clearly that the density of *Carex* is comparable with the density of the other species and that, when all species are considered together, the density on the grazed plot is nearly double that on the ungrazed plot.

8A.7 *Theoretical variances of transformed data*—Sometimes the residual variances of transformed data can be predicted by making assumptions concerning the distributions of the untransformed data. This can be done for the main transformations:

1 The square root transformation—If the observations are counts *e.g.* of bacteria, and if it is assumed that the objects counted are distributed at random*, the residual mean square tends to the value 0·25. If the average count is small *i.e.* less than five, the theoretical variance will lie somewhere between 0·0 and 0·4, but for an average count exceeding five the theoretical variance is nearly 0·25. Use of $\sqrt{(x+\frac{3}{8})}$ instead of $\sqrt{x}$ helps to stabilize

*This is equivalent to assuming that the counts are in Poisson distribution.

the variance and for average counts as low as two the theoretical variance is near 0·25. This is, in fact, demonstrated by the transformed worm egg counts of section 8.2. The estimated variance for the transformed counts using the transformation $\sqrt{x}$ tends to exceed the value 0·25, but the use of the transformation $\sqrt{(x+\frac{3}{8})}$ gives estimated variances nearer to this value.

In general, in any analysis the estimated variance should be used in preference to this theoretical value although it is sometimes permissible to use the theoretical variance. However, more usually the estimated variance should be tested against its theoretical value to ascertain whether the observations deviate from a purely random distribution. For example, the residual variance of the square roots of numbers of deaths in section 8.2 exceeded 1·0. It is obvious that the deaths cannot be considered as occurring at random. Evidently there is a tendency for the deaths to occur in particular litters.

2 The $\frac{1}{\beta}\sinh^{-1}\beta\sqrt{x}$ transformation—This is often required where counts are taken of objects which are not randomly distributed. If there is a tendency for the objects to be locally distributed at random with a varying density or to group together in randomly distributed groups the resulting (negative binomial) distribution should be transformed using $\frac{1}{\beta}\sinh^{-1}\beta\sqrt{x}$. The residual variance will then have a theoretical value 0·25 and the attainment of this value would indicate the underlying random distribution.

For example, this transformation applied to the sheep tick counts in section 8.2 gave a standard deviation of about 0·5 *i.e.* a variance of 0·25. Here this reflected the random distribution of counts on each sheep.

3 The $\sin^{-1}\sqrt{p}$ transformation—Here, if the proportions of objects with a particular trait which is randomly distributed are estimated in groups of size n*, the transformed proportions have a theoretical variance $0{\cdot}25/n$. Again the attainment of this value indicates the random distribution of the trait and a higher value indicates a tendency to group.

The percentages used in the example of section 8.2 were not based upon counts of discrete objects but, since they were estimated to the nearest 10 per cent, n might be taken as 10. For this value the theoretical variance is 0·025. The estimated variances all tend to exceed 0·025 but not very greatly. The variation between plots in this example was double that within plots. It should be concluded that although the distribution of species within each plot is nearly uniform there is a tendency towards grouping on particular plots.

*This is equivalent to assuming that the numbers with the trait are binomially distributed in each group.

8A.8 *Transformations of statistical measures*—Transformations of statistical measures, such as the correlation coefficient, may be employed to make them normally distributed and therefore capable of easier manipulation and comparison. The more common transformations of statistical measures are all based upon the logarithmic transformation. They will be considered in turn.

1 Estimated variances and standard deviations

The error in an estimated standard deviation is proportional to the standard deviation. Furthermore, interest centres in proportional, rather than absolute, changes in the standard deviation. These facts suggest use of the logarithm of the standard deviation or, calculated directly, half the logarithm of the variance. This transformed quantity will be denoted by z. Thus $z=\log s=0{\cdot}5 \log s^2$.

It may be shown that, provided the number of degrees of freedom n of the estimated variance is large (greater than 10), z tends to be normally distributed with variance $0{\cdot}0943/(n-1)$*. This fact makes the logarithm of the standard deviation a convenient quantity to consider and can be demonstrated by comparing *Figures* 37 and 49. The scatter of points in the former figure tends to increase with the mean, but no such tendency is observable in the latter. Any analysis, therefore, of the variation or changes in the standard deviation should be carried out using the logarithms of the standard deviations.

As an example consider the following set of variances calculated for observations made over a series of weeks. These variances, each with 41 degrees of freedom, were calculated using the logarithms of the (unpublished) worm egg counts taken by J. W. HOWIE:

Week	*1*	*2*	*3*	*4*	*5*	*6*
Variance	0·1610	0·1696	0·1418	0·2005	0·1782	0·2048
z	−0·897	−0·885	−0·924	−0·849	−0·875	−0·844
Week	*7*	*8*	*9*	*10*	*11*	*12*
Variance	0·1980	0·1280	0·2296	0·2527	0·1693	0·1861
z	−0·852	−0·946	−0·820	−0·799	−0·886	−0·865

There is apparently a tendency for the variance to increase over this period and a test might therefore be made to determine whether this increase is significant. If a regression of z upon time is carried out, the analysis of variance testing the significance of this regression is:

	D.f.	*S.s.*	*M.s.*	*V.r.*
Regression	1	0·00188	0·00188	1·10
Residual	10	0·01705	0·001705	
Total	11	0·01893	0·001721	

This shows that the regression sum of squares is no larger than might be expected by pure chance.

*If natural logarithms are used instead of logarithms to the base 10, this becomes $1/2\,(n-1)$.

The residual mean square in this analysis might be compared with the theoretical value $0{\cdot}0943/40 = 0{\cdot}0024$. It also is no larger (or smaller) than might be expected by pure chance so that it may be concluded that the variation in the estimated variances can be ascribed to chance.

This gives an approximate test of the uniformity or homogeneity of a set of variances. An alternative test will be given in section 8A.9.

II Variance ratios

Since the logarithm of a variance ratio is a difference between logarithms the methods of the previous section may be applied to the testing and comparison of variance ratios. Suppose s_1^2 and s_2^2 are two estimates of variance based upon n_1 and n_2 degrees of freedom and suppose $z = 0{\cdot}5 \log (s_1^2/s_2^2) = 0{\cdot}5 \log s_1^2 - 0{\cdot}5 \log s_2^2$. Then, for n_1 and n_2 large, z tends to be normally distributed with variance

$$0{\cdot}0943\left(\frac{1}{n_1 - 1} + \frac{1}{n_2 - 1}\right)^*$$

If there is no difference between the variances, then the mean value of z is zero. Consequently, by using the normal deviate table, it may be tested whether any variance ratio based upon large numbers of degrees of freedom is significant and, in addition, two variance ratios may be compared to determine whether one is significantly greater than the other. This latter test is, however, very approximate and the significance levels obtained by its use should be treated cautiously.

To test whether the standard deviations, 4·63 and 4·75, estimated in example 9 differ significantly from one another, z is calculated from

$$\log 4{\cdot}63 - \log 4{\cdot}75 = -0{\cdot}011$$

Its standard error is

$$\sqrt{[0{\cdot}0943\,(1/2050 + 1/2516)]} = \pm 0{\cdot}0091$$

The difference between the two standard deviations may thus easily be explained by chance variation.

III Correlation coefficients

The correlation coefficient may also be transformed and tested using the logarithmic transformation. Here, however, the transformation is

$$z = \frac{1}{2} \log \frac{1+r}{1-r}$$

This transformation is equivalent to the transformation for the variance ratio. The quantities

$$\frac{y}{\text{Standard deviation of } y} + \frac{x}{\text{Standard deviation of } x}$$

and

$$\frac{y}{\text{Standard deviation of } y} - \frac{x}{\text{Standard deviation of } x}$$

*As above, if natural logarithms are used this variance becomes $1/2(n_1 - 1) + 1/2(n_2 - 1)$. The quantity z is then called Fisher's z.

may be shown to be independently distributed. The variance of the former is $2+2r$ and the variance of the latter is $2-2r$. Thus their variance ratio is $(1+r)/(1-r)$.

The transformed quantity z is then approximately normally distributed with variance $0{\cdot}1886/(n-1)$, or $1/(n-1)$ if natural logarithms are used, where n is the number of degrees of freedom.

This transformation may be used to provide a test of a correlation coefficient, but it is more useful as a method of testing the difference between correlation coefficients or to obtain a combined estimate from a series of correlation coefficients.

To illustrate the use of this transformation consider two observed correlation coefficients between face length and face breadth for children 5 years old. For 66 boys the correlation coefficient was $-0{\cdot}009$ and for 60 girls the correlation was $0{\cdot}408$. To test the difference between these, first calculate the values of z. These are $-0{\cdot}0039$ and $0{\cdot}1881$. The difference between these values is $0{\cdot}1920$ and this has a standard error of $\sqrt{[0{\cdot}1886\,(1/63+1/57)]} = \pm 0{\cdot}0794$. The normal deviate testing the difference between the correlation coefficients is thus $0{\cdot}1920/0{\cdot}0794 = 2{\cdot}42$ and this value would occur by chance less than once in fifty times. There is consequently a strong indication of a facial difference between boys and girls at five years of age.

IV Estimated deviate t

The estimated deviate t may also be transformed to approximate normality. Here the appropriate transformation* is $z = \sinh^{-1}\sqrt{(t^2/n)}$ and the variance of z is $1/(n-1)$, where n is the number of degrees of freedom of t. This may be carried out easily, using *Table X* of the Appendix.

With this transformation values of t may be tested, two or more values of t may be combined, or, very approximately, it may be tested whether one value of t is significantly different from another.

Suppose there are two values of t, $2{\cdot}0$ and $1{\cdot}8$, with 20 and 15 degrees of freedom respectively. The corresponding values of z are $0{\cdot}43$ and $0{\cdot}44$ with variances $1/19$ and $1/14$. To test the two values together, calculate a combined estimate of z

$$\frac{0{\cdot}43 \times 19 + 0{\cdot}44 \times 14}{19+14} = 0{\cdot}434$$

This has a standard error of $\sqrt{[1/(19+14)]} = 0{\cdot}174$, so that the normal deviate testing the two values of t is $0{\cdot}434/0{\cdot}174 = 2{\cdot}49$. This would occur by chance about once in a hundred times so that it may be concluded that the two tests taken together are significant.

Alternatively, it will be more accurate if we transform the value of z back to t. This gives $t^2 = 34 \times 0{\cdot}204$ or $t = 2{\cdot}63$, with 34 degrees of freedom. As above, this is barely significant at the 1 per cent level.

*This may be shown to arise from the transformation of the last section if it is noted that

$$\frac{1+r}{1-r} = \left[\frac{t}{\sqrt{n}} + \sqrt{\left(1+\frac{t^2}{n}\right)}\right]^2$$

and

$$\ln\left[\frac{t}{\sqrt{n}} + \sqrt{\left(1+\frac{t^2}{n}\right)}\right] = \sinh^{-1}\sqrt{\left(\frac{t^2}{n}\right)}$$

8A.9 *Test for homogeneity of variance*—In order to test whether any transformation has been successful in stabilizing the variance, it is necessary to determine whether the set of transformed variances might have arisen from the same theoretical variance. The appropriate test for this has been given by BARTLETT, M. S. *Proc. roy. Soc., Lond., A* 160 (1937) 268. It may be stated as follows: If $s_1^2, s_2^2, \ldots s_k^2$ are k estimates of variance with n_1, $n_2, \ldots n_k$ degrees of freedom and s^2 is the pooled estimate of variance with n degrees of freedom, then, if these estimates are homogeneous, the value of

$$(n \log s^2 - n_1 \log s_1^2 - n_2 \log s_2^2 \; . \; . \; . \; - n_k \log s_k^2)/C$$

where

$$C = 0{\cdot}4343 \left[1 + \frac{1}{3(k-1)} \left(\frac{1}{n_1} + \frac{1}{n_2} + \ldots + \frac{1}{n_k} - \frac{1}{n} \right) \right]$$

is distributed approximately as a χ^2 with $k-1$ degrees of freedom. If natural logarithms are used the coefficient 0·4343 may be dropped.

In order now to test whether the variances of the transformed percentages of section 8.2 are homogeneous, the analysis proceeds as follows:

	Estimated variance s^2	*D.f.* n	$\log s^2$	$n \log s^2$	
	0·028	12	−1·553	−18·636	
	0·042	8	−1·377	−11·016	
	0·037	40	−1·432	−57·280	−369·604
	0·025	108	−1·602	−173·016	
	0·030	72	−1·523	−109·656	
Pooled	0·02922	240	−1·534	−368·160	

$$C = 0{\cdot}4343 \; [1 + 1/12 \, (1/12 + 1/8 + 1/40 + 1/108 + 1/72 - 1/240)]$$

$$= 0{\cdot}4343 \times 1{\cdot}021$$

$$= 0{\cdot}44342$$

$$\chi^2_{(4)} = \frac{-368{\cdot}160 + 369{\cdot}604}{0{\cdot}44342} = 3{\cdot}26$$

This value of $\chi^2_{(4)}$ is in close agreement ($P > 0{\cdot}5$) with what might be expected by chance. This shows that these variances may be considered as homogeneous and the pooled variance may in consequence be used.

It should be noted that in calculating C the portion in square brackets differed very little from 1·000. In fact, for rapid testing this bracketed expression may often be ignored and only introduced when there is some doubt about the significance. Any insignificant value obtained ignoring this expression cannot be made significant by its introduction.

8A.10 *Testing for normality*—It is also sometimes necessary to test whether the use of a transformation has normalized a non-normal distribution.

There are two methods by which this may be done. First, the mean and variance may be estimated and, using the table of the normal deviate, the numbers of observations falling within given limits may be calculated. These may then be tested against the observed numbers using the χ^2 test. Secondly, the observations may be used to estimate values which are known for the normal distribution. For example, the median for the normal distribution is equal to the mean. This latter approach requires a knowledge of the variability that is likely to occur in the estimated value.

To demonstrate the former approach consider the data of example 2. The estimated mean and standard deviation for the males are 0·5 and 20·3 respectively. Using *Table I*, the percentages of observations falling in the intervals, less than −30, −30- *etc* may be calculated as follows:

Grouping interval	*Observed* O	*Expected* E	$\frac{(O-E)^2}{E}$
Less than −30	7	6·7	0·01
−30-	8	9·0	0·11
−20-	15	14·6	0·01
−10-	22	18·8	0·55
0-	15	19·0	0·84
10-	18	15·1	0·56
20-	8	9·5	0·24
Over 30	7	7·3	0·01
	100	100·0	2·33

The expected and observed numbers may now be compared using the χ^2 test. Here χ^2 has 5 degrees of freedom since the mean and variance have both been estimated and the value 2·33 is insignificant. Evidently the normal distribution represents the observations reasonably well.

The most frequent application of the second method utilizes the moments defined in section 1A.13. There are two quantities which are more commonly used. If s^2 is the estimated variance and m_3 and m_4 the estimates of the third and fourth moments (based upon n observations), these are defined by

$$g_1 = \frac{n^2}{(n-1)(n-2)} \cdot \frac{m_3}{s^3}$$

and

$$g_2 = \frac{n^2(n+1)}{(n-1)(n-2)(n-3)} \cdot \frac{m_4}{s^4} - \frac{3(n-1)^2}{(n-2)(n-3)}$$

These rather complicated quantities reflect the skewness and kurtosis (or degree of flatness) respectively of the distribution that is being considered. For the normal distribution the average values of g_1 and g_2 will be zero so that non-zero values for g_1 and g_2 indicate departures from normality. In order to test whether any observed departures are significant or possibly due to chance it is necessary to know the variances of g_1 and g_2. These are:

$$\text{Variance of } g_1 = \frac{6n(n-1)}{(n-2)(n-1)(n+3)}$$

$$\text{Variance of } g_2 = \frac{24n(n-1)^2}{(n-3)(n-2)(n+3)(n+5)}$$

Alternatively, if n is large, these may be more rapidly estimated by taking the first few terms (usually only the first) of the expressions:

$$\text{Variance of } g_1 = \frac{6}{n+3}\left(1+\frac{2}{n^2}+\frac{2}{n^3}+\ldots\right)$$

$$\text{Variance of } g_2 = \frac{24}{n+5}\left(1+\frac{10}{n^2}+\frac{2}{n^3}+\ldots\right)$$

For instance, if $g_1 = -0{\cdot}21$, $g_2 = 0{\cdot}77$ and $n = 20$, these formulae give

$$\text{Variance of } g_1 = \frac{6}{23}\left(1+\frac{2}{400}+\frac{2}{8{,}000}+\ldots\right)$$

$$\text{Variance of } g_2 = \frac{24}{25}\left(1+\frac{10}{400}+\frac{2}{8{,}000}+\ldots\right)$$

The error in ignoring terms after the first is very small and, if this is done, the standard errors of g_1 and g_2 are $\sqrt{(6/23)} = 0{\cdot}51$ and $\sqrt{(24/25)} = 0{\cdot}98$ compared with the exact values of 0·51 and 0·99. Since neither of the estimates g_1 and g_2 exceeds its standard error, the above departure from normality is not significant.

Other tests of normality may be used if sufficient observations are available. If small numbers of observations are available, the above approach becomes very approximate. For the correct methods here, reference should be made to the work of R. C. GEARY *e.g. Biometrika* 34 (1947) 209.

8A.11 *Serial correlation*—One of the assumptions made in testing the difference between means or estimating the association between sets of measurements is that the observations in each set are independently distributed. If successive observations are not independent then the tests of significance described in the previous chapters will be invalid (although for the test of correlation coefficient to be invalid, successive observations in both sets have to be related). For this reason it is useful to have a test of whether successive observations are independent or not.

This test can be made by calculating the correlation coefficient between successive observations. Thus, if x_1, x_2, x_3, ... x_n are n successive observations, the correlation between the set of observations, $x_1, x_2, \ldots x_{n-1}$, and the observations immediately following these $x_2, x_3, \ldots x_n$ is calculated. This quantity is called a serial correlation coefficient. Since these two sets of $n-1$ observations are, in reality, only one set, the usual test of significance for a correlation coefficient has to be corrected in testing the serial correlation coefficient. This is done approximately by adding 2 degrees of freedom before testing the significance of the serial correlation coefficient. Hence the appropriate number of degrees of freedom for testing the correlation between $x_1, x_2, \ldots x_{n-1}$ and $x_2, x_3, \ldots x_n$ is $(n-1)-2+2=n-1$.

For example, the serial correlation coefficient for the monthly production of roofing slates for the years 1945 and 1946 was 0·667. Since the series here had 24 terms this correlation could be tested using the variance ratio

$$\frac{23(0{\cdot}667)^2}{1-(0{\cdot}667)^2}=18{\cdot}4$$

with 23 degrees of freedom. This value is highly significant. In consequence it is concluded that the production in successive months is highly correlated. This series of 24 items could not be correlated with other series unless some account was taken of the existence of the serial correlation. This might be done in various ways depending upon how the serial correlation arises; one of the simplest is to use a partial correlation coefficient eliminating the effect of the proceeding months.

The use of the serial correlation coefficient is not restricted to testing the dependence of successive observations; it may also be used to test the dependence of observations two apart, three apart and so on. However, the tests for time series are extensive and complicated and cannot be dealt with here.

SUMMARY OF PP 184 TO 194

Transformations to make effects additive have been given. The use of theoretical variances in testing transformed data using the square root, $\sin^{-1}\sqrt{p}$ or $\frac{1}{\beta}\sinh^{-1}\beta\sqrt{x}$ transformations have been demonstrated.

It has been shown how statistical measures, such as s^2, the variance ratio, r and t, may be transformed to normality and used as normal variates.

The tests for homogeneity of variance and normality have been given. Finally, the method of testing the interdependence of successive observations in a series of observations has been explained.

EXAMPLES

84 The following observations on antibody level were taken in two groups of sheep:

Group 1	2	30	5	3	20	2	7	7	20	20	50	2
	½	5	7	7	5	7	30	3	2	10	3	7
Group 2	1	7	5	7	1	2	30	5	7	5	3	20
	5	10	1	1	3	2	30	3	3	15	2	—

Why is the logarithmic transformation most suitable? Show that the use of this transformation stabilizes the variance and that the difference between the two groups is not significant.

85 The following figures give estimates of the percentage escape of haddock of different sizes from nets of commercial sized mesh.

Length cm	19	20	21	22	23	24
No. escaping	65	55	41	27	18	10

Transform the percentage escapes into normal deviates and use these to estimate the percentage escapes of haddock 25 cm long.

86 Test whether the variances of the tick counts of section 8.2 are homogeneous after the first week. Show that the transformed tick counts give rise to homogeneous variances. (*N.B.* Since the standard deviations are given, 2 log s should be used instead of log s^2.)

87 The production of coal (millions of tons) in Canada for the twelve years from 1937 to 1948 was 1·12, 1·00, 1·11, 1·25, 1·28, 1·33, 1·22, 1·18, 1·13, 1·23, 1·08 and 1·27. Show that the serial correlation coefficient is 0·32 and that this value is not significant.

88 1,000 observations of the right ascension of Polaris gave the following set of errors of observation.

Error in seconds	−3·5	−3·0	−2·5	−2·0	−1·5	−1·0	−0·5
Frequency	2	12	25	43	74	126	150
Error in seconds	0·0	0·5	1·0	1·5	2·0	2·5	3·0
Frequency	168	148	129	78	33	10	2

Obtain the following estimates: $s^2=1{\cdot}3386$, $m_3=-0{\cdot}3062$, $m_4=4{\cdot}8944$ and hence estimate $g_1=-0{\cdot}20$, $g_2=-0{\cdot}26$. Show that the standard errors of these values are 0·077 and 0·155 and hence that the skewness of this distribution is significant.

The skewness here is so small that it could in fact be ignored in making the usual tests based upon the assumption of normality. It is however of some interest in that it shows a tendency for the negative errors to be slightly larger than the positive errors.

89 Two experiments gave the values of t equal to 1·50 and 1·75 with 5 and 15 degrees of freedom respectively. Show that if these are combined using the χ^2 test for combination of probabilities the result is $P=0{\cdot}10$, while if the z transformation is used $P<0{\cdot}05$. This example again shows the advantage of being able to take the sign of the difference into account as well as the number of degrees of freedom in each value of t.

90 The correlations between weight and height of 66 boys and 60 girls each five years old were 0·670 and 0·737. Show that the difference between these coefficients is not significant and obtain the joint estimate 0·704 for the correlation between weight and height.

9

SAMPLING METHODS

9.1 *Random selection*—The application of statistical method depends largely upon the approach in which the chances against a set of observations arising naturally under a given hypothesis are calculated and if these chances are small the hypothesis is rejected. The idea of chance variation plays a large part in this approach. For example, if the accuracy of a mean is being determined, it is assumed that the individual observations constitute a reasonable selection of the whole. For this to be so the method of making the observations must be above criticism. As explained in section 1.1, any series of observations can be considered as a sample from a population of possible observations, so that the correct method of obtaining a sample set of observations is of some importance.

In sampling two main conditions should be satisfied: first, the sample should be unbiased and secondly, it should yield information on its own accuracy. These conditions have been adequately summarized by F. YATES who writes:

1 If bias is to be avoided, the selection of samples must be determined by some process uninfluenced by the qualities of the objects sampled and free from any element of choice on the part of the observer.

2 If a valid estimate of sampling error is to be available each batch of material must be so sampled that two or more sampling units are obtained from it. These sampling units must be a random selection from the whole aggregate of sampling units that can be taken from the batch of material, and all the sampling units in the aggregate must be of approximately the same size and pattern, and must together comprise the whole of the batch of material.

For the moment we shall consider the first problem: how to obtain an unbiased sample. This is not as easy at it appears, for the personal element intrudes. Conscious or unconscious selection by the observer occurs very easily, and where human beings are observed they often tend to influence their observers.

Many examples of bias may be found in the literature*, but the following serve to demonstrate two simple examples of a type of bias which is frequently encountered.

The first example consists of 200 tree girths measured by students. The students had been carefully instructed to take their results to the nearest inch. The following frequency table was obtained:

*For instance, YATES, F. Some Examples of Biased Sampling. *Annals of Eugenics* 6 (1935) 202

Girth	11	12	13	14	15	16	17	18	19	20	21	22
Frequency	~~2~~	~~6~~	5	10	9	12	15	16	20	23	~~17~~	~~12~~
Girth	23	24	25	26	27	28	29	30	31	32	33	*Total*
Frequency	10	17	7	6	1	6	1	2	0	~~2~~	1	200

If it is closely studied it will be seen that even numbers occur more frequently than their neighbouring odd numbers. This is shown more clearly if the table is presented according to the last figures of the girths:

Girths ending in	1	2	3	4	5	6	7	8	9	0
Frequency	~~19~~	~~20~~	18	27	16	18	16	22	21	25

Although there are only 112 even numbers against 88 odd numbers, the consistency of the 'see-saw' shows a distinct tendency for the even numbers to occur.

A second example is provided by a set of measurements of wool fibre widths. Here the measurements were taken to the nearest micron and they varied between 10 and 80 microns in width. The distribution of 1,000 last figures was:

Last figure	1	2	3	4	5	6	7	8	9	0
Frequency	86	110	66	126	79	109	74	124	78	148

Here there was an unmistakable tendency for 0 to occur and a lesser, but quite noticeable, tendency for even numbers to occur. This might be attributed to inaccurate rounding off but there is an indication that other effects are at work. The total number in the 1 and 9 classes is 164 compared with the expected number 200. The deficit of 36 must in part be accounted for by the excesses in the 2 and 8 classes and the remainder should account for the excess in the 0 class. This excess, which is 48, is however too large to be so accounted for.

Fortunately in these two examples these observed biases in the measurements are not likely to affect the conclusions appreciably. More important are the biases which cannot be easily discovered. Thus a tendency to under- or over-estimate a measurement cannot be easily discovered and may have disastrous effects. Alternatively a tendency to select for particular characteristics is liable to cause some distortion of the results. Thus the collection of economic data by calling from house to house is liable to give biased results unless houses at which no answer is received are subsequently revisited. Otherwise the houses with small families are likely to be under-represented since all the members of a small family are more likely to be out than all the members of a large family.

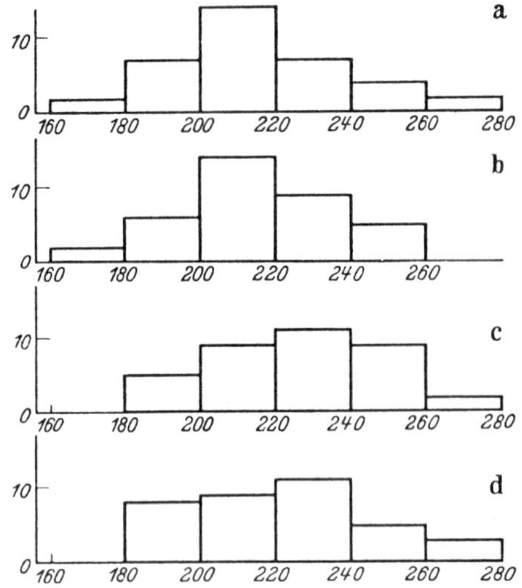

Figure 51. *Distributions of weaning weights*

The selection of particular characteristics may be completely unconscious but it may have disastrous results on the measurements or effects which are being estimated. For example, *Figure* 51 indicates the distributions of weaning weights in four groups of 36 rats allocated 'at random' for experimental purposes. The allocation here has resulted in the rats in the last two groups having weaning weights appreciably greater than the other two. This difference, which is highly significant, has apparently resulted from an unconscious selection by the experimenter.

Unless personal selection of this type can be ruled out, the estimation of means and effects is biased and standard errors derived for these estimates are meaningless. It is thus essential that such selection should be avoided and that the selection should be completely at random. This requires the use of a strictly impersonal method of random selection, such as can be achieved using a table of random numbers.

9.2 *Use of tables of random numbers*—In order to achieve unbiased selection a table of random numbers should be employed. Such tables, which are constructed so as to be free from bias, give series of randomly chosen numbers between 0 and 9. *Table VII* gives 2,000 such numbers and may be employed to achieve random selection, but for frequent use recourse should be had to one of the larger tables*. The various applications of this table in achieving random selection will now be considered.

Suppose it is required to select at random twelve individuals out of sixty. If the sixty are numbered in some manner, say according to the order in which they are encountered, it is required then to choose 12 numbers at random between 1 and 60. Reading down the first two columns of numbers in *Table VII*, we get: 34, 73, 98, 2, 10, 2, 47, 39, 45, 78, 42, 15, 94, 97, 44, 2, 97, 78, 18, 40, and so on. From these the numbers over sixty are rejected giving 34, 2, 10, 2, 47, 45, 42, 15, 44, 2, 18, 40, 21, 29, 9, and so on. Repeats

*FISHER, R. A. and YATES, F. *Statistical Tables for Biological, Agricultural and Medical Research* gives 15,000 random numbers. KENDALL, M. G. and BABINGTON-SMITH, B. have compiled a further 25,000 random numbers in *Tables of Random Sampling Numbers (Tracts for Computers No. 24)* Cambridge, 1939.

of any of these are also rejected (here 2 occurs three times) to give 34, 2, 10, 47, 45, 42, 15, 44, 18, 40, 21, 29 which supply the necessary 12 numbers. Of course, this process can be done directly and the numbers written down without any intermediate steps.

The same method can be applied to select at random from groups of any size, but sometimes it might require the rejection of a high proportion of the numbers. For example, if we want ten numbers between 1 and 20, by the above method we should select 2, 10, 15, 18, 9, 20, 1, 17, 4 and 19. This requires the use of the first four columns. It is in fact easier to divide twenty into each pair and use the remainder, 0 counting as 20. Here this would give 14, 13, 18, 2, 10, 7, 19, 5, 15, 17, and would use only the first seventeen pairs of numbers. However, in using this method it is essential to ensure that each number has an equal chance of being represented. Thus, if ten numbers between 1 and 21 had been required the remainder might still have been used, but numbers 85-99 and 00 should be rejected. This uses only the numbers 1-84 and gives each number between 1 and 21 an equal chance of occurring. The numbers would then be 13, 10, 2, 5, 18, 3, 15, 21, 19, and 16.

A last method which avoids lengthy division is to use in effect only the first figure in division. Thus, if 10 numbers between 1 and 21 are required, numbers are first selected at random between 1 and 30 and then those between 22 and 30 are rejected. To do this thirty is divided into the first two figures, rejecting 91-99 and 00, to give the remainders 4, 13, 2, 10, 2, 17, 9, 15, 18, 12, 15, 14, 2, 18, 18, 20. . . . Rejection of repeats and numbers over 21 then gives 4, 13, 2, 10, 17, 9, 15, 12, 14 as the ten random numbers. As a second example of this method suppose 10 numbers between 1 and 150 are required. Using the first three columns these would be 148, 139, 27, 105, 22, 72, 58, 20, 147, and 41.

A second type of problem which can be solved using a table of random numbers is to arrange a series in random order. This is done by allocating a number to each number of the series and then selecting a complete set at random. For example, if a set of 8 objects is to be arranged in random order, 8 numbers must be chosen at random between 1 and 8. Using the first column in *Table VII* and rejecting 0 and 9 the resulting selection is 3, 7, 1, 4, 2, 8, 5, and 6.

This may be rather tedious if a large series has to be arranged in order and it is easier here to carry out the selection in two or more stages. For example, suppose the numbers between 1 and 40 are to be rearranged in random order. Using the eleventh and twelfth columns of the first page of *Table VII* the first twenty numbers are: 28, 24, 21, 8, 35, 20, 31, 11, 23, 4, 30, 2, 33, 36, 1, 19, 32, 17, 37, 5. Repetition of numbers now makes further selection difficult so that the numbers already used may be struck out and numbers between 1 and 20 chosen. The remaining numbers are:

~~1~~	~~2~~	3	~~4~~	~~5~~	6	7	~~8~~	9	10
~~11~~	12	13	14	15	16	~~17~~	18	~~19~~	~~20~~
~~21~~	22	~~23~~	~~24~~	25	26	27	~~28~~	29	~~30~~
~~31~~	~~32~~	~~33~~	34	~~35~~	~~36~~	~~37~~	38	39	40

and using the sixteenth and seventeenth columns we get 10, 1, 12, 15, 2, 6, 7, 5, 13, 4. The tenth, first . . . fourth numbers in the above array are 16, 3, 22, 27, 6, 12, 13, 10, 25, 9, and when these are struck out we get:

~~1~~	~~2~~	~~3~~	~~4~~	~~5~~	~~6~~	7	~~8~~	~~9~~	~~10~~
~~11~~	~~12~~	~~13~~	14	15	~~16~~	~~17~~	18	~~19~~	~~20~~
~~21~~	~~22~~	~~23~~	~~24~~	~~25~~	26	~~27~~	~~28~~	29	~~30~~
~~31~~	~~32~~	~~33~~	34	~~35~~	~~36~~	~~37~~	38	39	40

The order for the remaining ten numbers, as determined from the eighteenth column, is 10, 1, 6, 3, 4, 7, 9, 8, 2, 5. In the above array these give the numbers 40, 7, 29, 15, 18, 34, 39, 38, 14, 26. The completed ordering is thus:

28	24	21	8	35	20	31	11	23	4
30	2	33	36	1	19	32	17	37	5
16	3	22	27	6	12	13	10	25	9
40	7	29	15	18	34	39	38	14	26

9.3 *Randomization in experimentation*—In carrying out experiments it is always necessary to ensure that the treatments are randomly arranged within the framework of the experiment. This may be done by the methods described in the previous section, but it is necessary to ensure that a completely random arrangement is achieved. To do this, it is necessary in general to carry out three steps:

1 a design of the type to be used should be chosen at random from the possible designs of this type

2 treatments should be allotted at random to the treatment letters of the design

3 the rows, columns and blocks of the design should be randomized.

Sometimes one or more of these steps may be unnecessary *e.g.* in using randomized blocks, the second and third steps are unnecessary if the first step is correctly carried out. Usually, if step *1* can be perfectly achieved step *2* can be ignored, but often the designs are tabulated in some systematic fashion and step *2* must therefore be carried out. For example, most tabulations of lattice square designs have treatments 1-5 occurring in the same row or column. It might be disastrous, however, to allocate these to the first five treatments, which are often the lower levels of the treatments tested.

Randomization of any particular design may be carried out as described in the previous section. As an example consider the randomization of a 6 × 6 Latin square. One was selected at random from FISHER, R. A. and YATES, F. *Statistical Tables for Biological, Agricultural and Medical Research*:

A	*B*	*C*	*D*	*E*	*F*
B	*A*	*F*	*E*	*C*	*D*
C	*F*	*B*	*A*	*D*	*E*
D	*C*	*E*	*B*	*F*	*A*
E	*D*	*A*	*F*	*B*	*C*
F	*E*	*D*	*E*	*A*	*B*

Using the first row of the second page of *Table VII*, the columns may be rearranged at random in the order 5, 3, 4, 1, 6, 2. This gives the square:

E	*C*	*D*	*A*	*F*	*B*
C	*F*	*E*	*B*	*D*	*A*
D	*B*	*A*	*C*	*E*	*F*
F	*E*	*B*	*D*	*A*	*C*
B	*A*	*F*	*E*	*C*	*D*
A	*D*	*E*	*F*	*B*	*E*

Next the rows have to be rearranged at random. From the second row of the second page of *Table VII*, the order 1, 2, 5, 6, 4, 3 is obtained. The rearranged square is then:

E	*C*	*D*	*A*	*F*	*B*
C	*F*	*E*	*B*	*D*	*A*
B	*A*	*F*	*E*	*C*	*D*
A	*D*	*E*	*F*	*B*	*E*
F	*E*	*B*	*D*	*A*	*C*
D	*B*	*A*	*C*	*E*	*F*

Lastly, the treatments are allocated at random to the letters. Using the third row of the second page of *Table VII*, the arrangement is: *A*, 4; *B*, 5; *C*, 1; *D*, 2; *E*, 6; *F*, 3. Thus *C* is used to denote the first treatment, *D*, the second, and so on. This completes the randomization.

9.4 *Methods of sampling*—Now consider general methods of sampling which satisfy the requirements stated in section 9.1. The sample must be unbiased and yield an estimate of its own error but, consistent with these aims, the sample should be representative *i.e.* it should have nearly the same characteristics as the population. For this reason, purely random sampling is seldom employed. Two main alternatives exist: stratified random sampling and systematic sampling.

In stratified random sampling, the material or area to be sampled is divided into strata or groups and a number of observations is taken from each stratum. To yield an estimate of the accuracy, in general at least two samples are required from each stratum. The strata should be chosen to be as uniform as possible so that by ensuring that each stratum is proportionally represented in the total sample the variability between strata is eliminated. For instance, if the sampling is to determine the average annual expenditure of families on different articles, the area concerned may be divided into comparable districts and a proportion of families taken from each district. In this manner each district is proportionally represented in the total sample.

Sometimes, however, the population to be sampled may be divided into strata some of which are more variable than others. For example, we may stratify the population of a town according to age and decide to take a proportion from each age group. If, however, the measurements or observations that are being taken are more variable in some age groups it may be desirable to take a greater proportion from these age groups to determine these groups more accurately. Such a procedure is known as stratified sampling with a variable sampling fraction.

Ideally, where this method is employed, the proportion taken from each stratum should be roughly proportional to the standard deviation within each stratum. Thus, if one stratum has twice as large a standard deviation as another, the proportion sampled in the first stratum should be roughly double that in the second.

In systematic sampling the samples are equally spaced throughout the area or population to be sampled. Thus, for example, in house-to-house sampling every tenth or twentieth house may be taken, or in sampling a field a sample may be taken every fifteenth pace. By this means we may ensure that the whole population is represented in the final sample. There are, however, several analytical difficulties connected with systematic sampling. First, it is necessary that a systematic sample should have a starting point chosen at random. If every tenth house is being taken, then a house would have to be chosen at random from the first ten as a starting point. Secondly, in order to obtain a valid estimate of the accuracy of the sample, it is necessary to take at least two systematic samples. To improve the estimate of error the material might be divided into strata and independent systematic samples taken from each. Alternatively, if a single systematic sample is taken, it may be used to obtain an overestimate for the error *i.e.* it may be employed to give a lower limit for the accuracy of the sample. The method of doing this will be explained in the section 9.6.

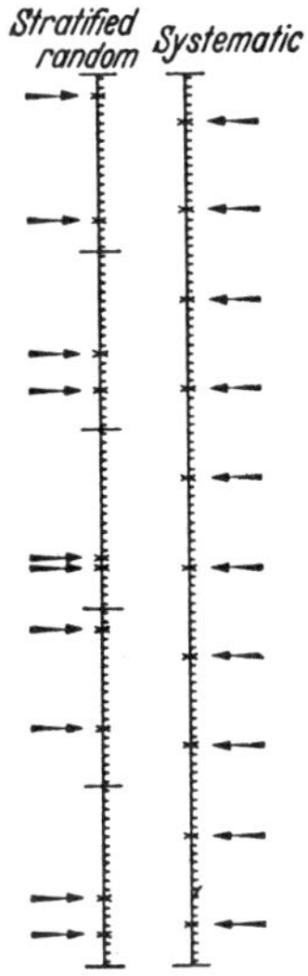

Figure 52
Methods of sampling

Figure 52 gives examples of stratified random and systematic sampling along a line. For the stratified random sample the line is divided into equal strata of twenty units and two samples (indicated by crosses) are taken at random from each. For the systematic scheme samples are taken at intervals of ten units, the starting point being chosen at random. A similar figure might be constructed for samples taken over an area. Here the systematic sample would appear as a grid of samples, while for the stratified random sample the area would be divided into sub-areas and a number of samples taken at random from each sub-area.

9.5 *Analysis of stratified random samples*—The form of analysis adopted for the examination of results depends upon the method of sampling used: stratified random or systematic. For stratified random sampling, the method of analysis will also depend upon whether a fixed or variable proportion is taken from each stratum. For simplicity suppose that we are sampling for one variable only and that where the population is stratified the size of each stratum is known.

Suppose that the sizes of the k strata are $s_1, s_2, \ldots s_k$, that $n_1, n_2, \ldots n_k$ observations are taken from these strata, that the standard deviations within the strata are $\sigma_1, \sigma_2, \ldots \sigma_k$ and that the estimated means from the strata are $\bar{x}_1, \bar{x}_2, \ldots \bar{x}_k$. Also suppose that the total size of population sampled is S, the total number of observations taken is N, and that the estimated mean of the observation is $\bar{x}$.

If a fixed proportion is sampled from each stratum, then

$$\frac{n_1}{s_1} = \frac{n_2}{s_2} = \ldots = \frac{n_k}{s_k} = \frac{N}{S} = \frac{\text{No. sampled from each stratum}}{\text{Size of stratum}}$$

and the overall mean $\bar{x}$ is the best estimate for the whole population. The standard error of this (using the rule of section 3.5) is

$$\sqrt{\left[\frac{n_1\sigma_1^2 + n_2\sigma_2^2 + \ldots + n_k\sigma_k^2}{N^2}\right]}$$

Usually, of course, if the standard deviations were different, a variable sampling fraction would be used, so that if a fixed sampling fraction is used $\sigma_1, \sigma_2, \ldots \sigma_k$ may generally be replaced by σ. The standard error then becomes $\sigma/\sqrt{N}$ as usual. Here the form of analysis is exactly the same as that used in testing the differences between several groups, the differences between strata being eliminated in the same manner and the residual mean square being used as an estimate of σ^2.

If the standard deviations are the same in all strata but the numbers taken from each stratum are not proportional to the stratum size, then the overall mean should be estimated from

$$\frac{s_1\bar{x}_1 + s_2\bar{x}_2 + \ . \ . \ . \ + s_k\bar{x}_k}{S}$$

and the standard error of this estimate is

$$\frac{\sigma}{S}\sqrt{\left[\frac{s_1^2}{n_1} + \frac{s_2^2}{n_2} + \ . \ . \ . \ + \frac{s_k^2}{n_k}\right]}$$

Here again σ^2 is estimated using the residual mean square in the analysis of variance after eliminating differences between strata.

If the proportion in each stratum is proportional to the standard deviation within the stratum, then

$$\frac{n_1}{s_1\sigma_1} = \frac{n_2}{s_2\sigma_2} = \ . \ . \ . \ = \frac{n_k}{s_k\sigma_k} = \frac{N}{s_1\sigma_1 + s_2\sigma_2 + \ . \ . \ . \ + s_k\sigma_k}$$

The best overall estimate for the population is still given by

$$\frac{s_1\bar{x}_1 + s_2\bar{x}_2 + \ . \ . \ . \ + s_k\bar{x}_k}{S}$$

and its standard error is now

$$\frac{\sqrt{N}}{n_1/\sigma_1 + n_2/\sigma_2 + \ . \ . \ . \ + n_k/\sigma_k}$$

Lastly, if the proportions taken from each stratum are not exactly proportional to the standard deviations, the above estimate may still be used but the standard error becomes

$$\frac{1}{S}\sqrt{\left[\frac{s^2{}_1\sigma_1{}^2}{n_1} + \frac{s^2{}_2\sigma_2{}^2}{n_2} + \ . \ . \ . \ + \frac{s^2{}_k\sigma_k{}^2}{n_k}\right]}$$

To demonstrate the uses of these formulae we shall consider sampling from a population with three strata of relative sizes 1 : 2 : 3 and with standard deviations 4, 3 and 2 respectively *i.e.* $s_1 = 1$, $s_2 = 2$, $s_3 = 3$, $\sigma_1 = 4$, $\sigma_2 = 3$, $\sigma_3 = 2$, and suppose that 96 observations in all are taken. (This type of population in which the smaller strata are more variable is not uncommon.)

If a fixed sampling fraction is used, then

$$\frac{n_1}{1} = \frac{n_2}{2} = \frac{n_3}{3} = \frac{96}{6}$$

i.e. $n_1 = 16$, $n_2 = 32$, $n_3 = 48$. With these numbers sampled from each stratum the mean of the 96 observations estimates the overall mean. This estimate has a standard error of

$$\sqrt{\left[\frac{16 \times 4^2 + 32 \times 3^2 + 48 \times 2^2}{96^2}\right]} = \pm 0{\cdot}283$$

Alternatively, if a variable sampling fraction is used, then the numbers to be taken from each stratum are determined by:

$$\frac{n_1}{4} = \frac{n_2}{6} = \frac{n_3}{6} = \frac{96}{4+6+6}$$

i.e. $n_1 = 24$, $n_2 = 36$, $n_3 = 36$. With these numbers in each stratum the mean is estimated from

$$\frac{\bar{x}_1 + 2\bar{x}_2 + 3\bar{x}_3}{6}$$

This has a standard error of

$$\frac{\sqrt{96}}{24/4 + 36/3 + 36/2} = \pm 0{\cdot}272$$

This value is not very much smaller than that obtained by taking a fixed proportion from each strata. A larger difference between the standard deviations would be required to make the variable sampling fraction profitable.

Of course, in general, the standard deviations are not known, so that the appropriate numbers in each group can be gauged only approximately. If, in the above example, the standard deviations were incorrectly estimated as 3·5, 3·4 and 1·9, the required numbers in each group would be estimated as $n_1 = 21$, $n_2 = 41$, $n_3 = 34$. The overall mean would still be estimated from

$$\frac{\bar{x}_1 + 2\bar{x}_2 + 3\bar{x}_3}{6}$$

and its standard error would be

$$\frac{1}{6}\sqrt{\left[\frac{1^2 \times 4^2}{21} + \frac{2^2 \times 3^2}{41} + \frac{3^2 \times 2^2}{34}\right]} = \pm 0{\cdot}274$$

This differs to a negligible degree from the values obtained above using exact sampling fractions. Evidently, the accuracy is not very greatly affected by small inaccuracies in estimating the standard deviations.

For practical field sampling, the easiest method of carrying out the sampling is to divide the area into k equally sized strata and to take two samples from each stratum. The mean of these samples then estimates the overall mean. If $d_1, d_2, \ldots d_k$ are the differences between the pairs of samples in each stratum the standard error of this mean is estimated from

$$(1/2k)\sqrt{(d_1^2+d_2^2+\ldots+d_k^2)}$$

with k degrees of freedom.

For example, if 2, 4; 5, 6; 5, 3; 4, 4; 3, 1; 1, 0 are six such pairs of samples the estimated mean is $(2+4+\ldots+0)/12=3{\cdot}17$ and its standard error is

$$(1/12)\sqrt{(2^2+1^2+2^2+0^2+2^2+1^2)} = \pm 0{\cdot}312$$

with 6 degrees of freedom.

This method is most useful when the variability in each stratum is roughly the same.

9.6 *Analysis of systematic samples*—The analysis of systematic samples presents some difficulty since they are liable to be affected by any trend in the observations. Thus each observation in the series $x_2, x_{12}, x_{22}, x_{32}, \ldots$ is one unit behind the corresponding observation in the sample $x_1, x_{11}, x_{21}, x_{31}, \ldots$ and in consequence, if there is an upward trend in the observations, the mean of the former sample will be larger than that of the latter sample. The effect of such a trend would be even more marked if extreme samples such as $x_{10}, x_{20}, x_{30}, \ldots$ are considered.

A method of adjusting systematic samples* has been suggested to overcome this effect when there is a strong trend but, since this method might not correct the bias completely, it should be used cautiously.

An estimate of error may be obtained by taking several systematic samples and, treating each as a unit, estimating the variance between them. If the population is stratified and systematic samples are taken in each stratum, then each systematic sample may be considered as a unit and the whole sample analysed like a stratified random sample.

Alternatively, an upper limit for the error may be obtained by calculating the mean squared difference between successive samples. If this is denoted by d^2 and if n samples are taken, a rough estimate of the standard error is

*YATES, F. Systematic sampling. *Phil. Trans. roy. Soc., Lond. A* 241 (1948) 345

given by $d/\sqrt{2n}$. For example, if 2, 4, 5, 6, 5, 3, 4, 4, 3, 1, 1, 0 is a series of equally spaced samples, the value of d is

$$\sqrt{\left[\frac{1}{11}\left(2^2+1^2+1^2+1^2+2^2+1^2+0^2+1^2+2^2+0^2+1^2\right)\right]} = \pm 1{\cdot}279$$

and the approximate standard error is

$$\frac{1{\cdot}279}{\sqrt{24}} = \pm 0{\cdot}261$$

The use of systematic sampling is not generally to be recommended unless very large samples are to be taken and a great deal of care is to be expended in the analysis of them.

9.7 *Sampling from finite populations*—Specification of the accuracy of a mean by a standard error was introduced in Chapter 3 as a method of indicating the limits within which the true mean was likely to vary. The true mean here was regarded as a value which could only be realized exactly by taking an infinite number of observations. For example, in estimating a physical constant, such as Joule's, its estimate may be improved by taking more observations, but to determine it exactly an infinite number of observations would be required. This is reflected in the standard error which takes the value zero only for an infinite number of observations.

Sometimes, however, by taking sufficient observations it is possible to determine a mean exactly. Thus, if every member of a population is observed, as at a census or any complete enumeration, the mean may be determined exactly. It will be generally true that if we wish to measure exactly the mean of a finite population we shall be able to do so by taking sufficient observations. This means that the formula used for the standard error of a mean will be incorrect if a finite population is being considered.

Usually, even when the population is finite, it is so large that the formula for the standard error of the estimated mean is very accurate. For instance, if an attempt is made to estimate the mean length of haddock in the North Sea, or public opinion in Great Britain, the relevant populations are so large that a sample of over a million would be required before the effect of their finite size could be felt. For the usual size of sample the finite nature of such a population may be neglected. On some occasions the sample may, however, represent a fairly high proportion of the total population. Then it is necessary to acknowledge the finite size of the population in estimating the standard error.

It has been pointed out that if all of a population is observed the mean is known exactly and its standard error must be zero, but if a fraction f of the total population is sampled the estimated mean will in general deviate from the true mean of the entire population. The estimate of the true mean may then be considered as being made up of two parts: the estimate of the fraction f, which is known exactly and consequently has a zero variance, and the estimate of the unknown fraction $(1-f)$. This latter estimate must be based upon the former, but the variance of the estimated mean of the unknown fraction will not be influenced by any consideration of the finite population *i.e.* it will be σ^2/n. The variance of the two portions combined *i.e.* of the estimated mean, will then be

$$f \times 0 + (1-f) \times \frac{\sigma^2}{n} = \frac{(1-f)\,\sigma^2}{n}$$

Thus the variance is altered by the factor $(1-f)$, and if only a small fraction of the population is sampled it will be reduced to σ^2/n.

An alternative derivation of this formula is worth noting. The finite population itself may be considered as being a sample from a hypothetical infinite population; then

$$\begin{array}{c}\text{Variance of sample mean}\\ \text{about infinite population}\\ \text{mean}\end{array} = \begin{array}{c}\text{Variance of sample}\\ \text{mean about finite}\\ \text{population mean}\end{array} + \begin{array}{c}\text{Variance of finite popula-}\\ \text{tion mean about infinite}\\ \text{population mean}\end{array}$$

If the finite population consists of N members of which n are in the sample *i.e.* a fraction $f=n/N$ is sampled, then this becomes:

$$\frac{\sigma^2}{n} = \begin{array}{c}\text{Variance of sample}\\ \text{mean about finite}\\ \text{population mean}\end{array} + \frac{\sigma^2}{N}$$

i.e.

$$\begin{array}{c}\text{Variance of sample}\\ \text{mean about finite}\\ \text{population mean}\end{array} = \frac{\sigma^2}{n} - \frac{\sigma^2}{N} = \frac{\sigma^2}{n}\left(1 - \frac{n}{N}\right) = \frac{(1-f)\sigma^2}{n}$$

as previously.

This extra term is easily included in determining the standard error of an estimated mean, but it is seldom required in calculating the standard error of the difference between means as we are usually interested in drawing conclusions which are generally applicable and not restricted only to the finite populations which are being sampled. If, however, we are concerned only with the finite population, the factor $(1-f)$ should be included in calculating the standard errors of differences.

As an example of the application of this factor consider the data of example *1*. Here 460 tree girths were measured out of a total stand four or five times as large. The mean girth 44·51 in estimates the mean of the whole stand and, since f lies between 1/4 and 1/5, its standard error lies between

$$10{\cdot}78\sqrt{\left[\frac{1}{460}\left(1-\frac{1}{4}\right)\right]}=0{\cdot}44$$

and

$$10{\cdot}78\sqrt{\left[\frac{1}{460}\left(1-\frac{1}{5}\right)\right]}=0{\cdot}45$$

It should be observed that the standard error is fairly insensitive to the value of f, so that even when the total size of the population is not known a rough estimate will often be sufficiently accurate.

This estimated standard error should be taken as the larger of these values *i.e.* 0·45, and this could be used to compare the timber in this particular stand with the timber in any other stand to test whether one contains more than the other. If, however, the comparison with another stand was being made to answer a wider question, such as whether the method of cultivation or thinning applied to one stand is better than that applied to another, the estimated standard error must be calculated from $\sigma/\sqrt{n}$, since a hypothetical infinite population is being considered, of which this is only a sample. Furthermore, this estimate of σ^2 cannot be derived from the variation within any particular stand, but must be estimated from the variation between similar stands.

As a second example consider the sampling of attributes in a finite population. Suppose, of a group of 1,000 people, 400 are sampled at random and of these 120 possess a certain attribute, say, for example, they hold a certain opinion. Then the estimated proportion with that opinion in the whole group is 120/400=0·30 and this, now, has a standard error of

$$\sqrt{\left[\frac{0{\cdot}30\,(1-0{\cdot}30)}{400}\times\left(1-\frac{400}{1{,}000}\right)\right]}=+0{\cdot}018$$

Thus, with 95 per cent certainty, the true percentage in the group lies between 0·25 and 0·35. The high proportion sampled here leads to an appreciable reduction in the standard error. If no account were taken of the finite population, this would be

$$\sqrt{\left[\frac{0{\cdot}30\,(1-0{\cdot}30)}{400}\right]}=+0{\cdot}023$$

Again, the former standard error should only be used to compare the percentage in this group with another percentage if the results so obtained are to be considered as referring to this group alone. If, however, results of general applicability are required, the latter standard error has to be used.

9.8 *Sampling efficiency*—The relative efficiencies of different methods of sampling may be determined using the variances of the means derived by the different methods. For any particular method (with a large population) the effect of doubling the size of sample is to halve the variance of the estimated mean. Thus, if the variance of a mean estimated by one method of sampling is double that of a mean estimated by a second method of sampling using a sample of the same size, the second method is said to be twice as efficient as the first since it produces with a sample of, say, 100 a mean as accurate as would be produced by the first method with a sample of 200. In this manner the reciprocal variances (sometimes called invariances) of the means estimated by different methods of sampling reflect the relative efficiencies of the methods. It is often convenient to adjust the reciprocal variance of purely random sampling to take the value 100. The values taken by other methods of sampling then indicate the number of observations required in purely random sampling to obtain the same accuracy as could be obtained by 100 observations using the particular method of sampling that is being considered.

For example, if 2, 4, 5, 6, 5, 3, 4, 4, 3, 1, 1, 0 is a random sample, the estimated variance is

$$\frac{1}{11}\left(2^2+4^2+\ .\ .\ .\ +0^2-\frac{38^2}{12}\right)=3{\cdot}424$$

and the standard error of the estimated mean is

$$\sqrt{\left(\frac{3{\cdot}424}{12}\right)}=\pm 0{\cdot}534$$

This might now be compared with the standard error 0·312 obtained for the stratified random sample of 12 used in section 9.5 and with the approximate standard error ±0·261 obtained for the systematic sample of 12 in section 9.6. The reciprocal variances here are 3·507 for the purely random sample, 10·273 for the stratified random sample and 14·680 for the systematic sample. If the random sample is adjusted to 100, these become 100, 293 and 419, showing that about 300 random observations are required to give as accurate an estimate as 100 stratified random observations and about 400 random observations are required to give as accurate an estimate as 100 systematic observations.

These figures refer, of course, only to this particular example and should not be regarded as being generally applicable. The relative efficiencies of the three methods of sampling will depend upon the form of population sampled, but if the population can be broken up easily into fairly uniform strata (as in this example) relative efficiencies of the above order of magnitude might be expected. In comparing the relative efficiencies of different methods of sampling the question of cost, time and labour involved must also be considered. If one method of sampling is twice as efficient as another, but requires three times as much work to carry it out, it cannot be recommended for general use. To obtain the best method of sampling we have, in fact, to consider efficiency per unit cost in work, time and money. Such a consideration is beyond the scope of this book and the interested reader should refer to a more specialized text for information*.

SUMMARY OF PP 196 TO 211

Methods of random selection avoiding bias have been considered and, in particular, the uses of a table of random numbers have been demonstrated.

Different methods of sampling have been given and the methods of analysing sample results shown. It has also been shown how the standard errors of the estimated means of finite populations may be calculated.

Finally, it has been shown how, using the reciprocal of the variance, relative efficiencies of different methods of sampling may be compared.

EXAMPLES

91 The frequencies with which the digits 0-9 occur in *Table VII* are:

Digit	0	1	2	3	4	5	6	7	8	9
Frequency	204	204	202	196	221	209	185	194	189	196

Show that these frequencies are well within the normal limits of variation.

92 Using *Table VIII*, take a random sample of 12 logarithms to base ten between 1·00 and 9·98 and use these to estimate the mean logarithm in this range.

Dividing the total range into the three strata 1·00 – 3·98, 4·00 – 6·98, 7·00 – 9·98, take a stratified random sample with 4 samples from each stratum.

Finally, take a stratified random sample with a variable sampling fraction from the above strata. Here the standard deviations may be gauged roughly from the ranges within the three strata: 0·60, 0·24, 0·16. Thus 7, 3 and 2 samples respectively may be taken from the three strata.

The true mean is in fact 0·6757 and these three methods should give, on average, successively more accurate estimates of this value The standard errors of these estimates are approximately $\pm 0{\cdot}0739$, $\pm 0{\cdot}0310$ and $\pm 0{\cdot}0271$.

93 Using the standard errors given in the previous example show that the relative efficiencies of the three methods of sampling used are roughly 100 : 568 : 744. 0.

94 The sugar beets on a 1-acre field were sampled for sugar percentage. This was done by dividing the field into fifteen strata, each of four rows. The sugar percentage was then estimated for two samples of 10 beets from each stratum with the following results:

*Yates, F. *Sampling Methods for Censuses and Surveys* London, 1949, for example.

Stratum	*1*	*2*	*3*	*4*	*5*	*6*	*7*	*8*
Sample 1	14·58	13·35	13·90	13·49	14·92	14·71	14·48	15·01
Sample 2	13·81	13·87	14·31	14·78	14·14	13·44	14·85	14·58
Stratum	*9*	*10*	*11*	*12*	*13*	*14*	*15*	*Total*
Sample 1	14·28	14·38	14·14	14·38	15·04	14·58	16·20	432·48
Sample 2	14·24	14·46	13·73	14·27	15·19	14·87	14·50	

This constituted a 1 in 20 sample of the whole field.

Show that the mean sugar percentage of the whole crop is $14{\cdot}42 \pm 0{\cdot}094$ (the actual figure for the whole crop was 14·52).

95 The total number of 18-year old Aberdeen students in 1947 was 293. Using this information show that the limits in example *22* may be changed to 143·94 and 149·38 lb.

96 In a sample of 200 taken from a population of 980, five members are observed with a given attribute. Show that the proportion with the attribute is $0{\cdot}025 \pm 0{\cdot}0098$.

97 In a population of 8,000 roughly one half held a certain opinion. Show that a sample of approximately 1,330 would be required to determine the true proportion to within 0·025 with 95 per cent certainty. Find the size of sample required to reduce this to 0·01 with 95 per cent certainty (Answer: 4,500).

98 The following table gives an observed distribution of arm length in mm of the brittle-star *Ophiocoma nigra*:

Final Digit \ *Length mm*	20+	30+	40+	50+	60+	70+	80+
0	2	3	18	32	28	10	0
1	2	9	14	48	27	6	1
2	3	11	26	32	27	3	2
3	3	12	22	43	15	4	—
4	3	16	20	41	12	7	—
5	4	5	26	28	12	4	—
6	7	13	23	31	15	2	—
7	2	17	31	31	17	1	—
8	13	29	64	32	19	0	—
9	4	14	21	26	6	1	—

Show that there is a bias towards 8's and away from 9's.

99 The following table gives the distribution of final digits in head measurements of male and female inmates of Scottish asylums in 1903.

Digit	*Males*		*Females*	
	Breadth	*Length*	*Breadth*	*Length*
0	457	441	385	401
1	429	429	391	399
2	465	467	397	402
3	455	462	394	406
4	427	427	401	387
5	440	453	412	391
6	431	418	401	374
7	429	422	362	391
8	462	477	420	410
9	441	440	388	390

Construct a table showing the differences between the frequencies of occurrence of each digit and the mean frequencies of the two adjacent digits. Hence show that there is an excess of 8's and a deficiency of 7's.

EXTENDED DEVELOPMENT

9A.9 *Multi-stage sampling*—In large scale sampling surveys the procedure is generally more complicated than has been indicated in previous sections. The problem of bias in the observer and observed becomes more acute and great care has to be exercised in the taking of samples. The methods of sampling, while retaining the basic ideas expressed above, become more involved and, correspondingly, analysis of the samples is more complicated. In the remainder of this chapter the ideas behind some of these methods will be explained, but it is not intended to give a complete account of their operation. The bibliography should be referred to for references to particular aspects of these methods.

One of the chief difficulties in large scale surveys is the extensive area that may have to be covered in getting a random or stratified random sample. It may be a very expensive and lengthy task to cover the whole population in order to obtain a representative or unbiased sample. It would, in fact, be much better if it were possible to cover a small portion of the population intensively and thus obtain an unbiased estimate.

This can, in fact, be done provided that the portion to be studied is randomly chosen and also provided that some idea can be obtained of the sampling error resulting from such a choice. The sampling procedure here is to break the population up into groups, to take a sample (random or stratified random) of the groups and then to take samples from within each of the chosen groups. Thus the sampling would be carried out in two stages: the first stage determining the groups, and the second stage sampling the groups. Correspondingly, the error in the estimated mean can be split into two portions: the portion arising from the variation between the groups and the portion arising from the variation within the groups.

If there is no variation within the groups the variance of the estimated mean is

$$\frac{\text{Variance between group means}}{\text{Number of groups sampled}} = \frac{\sigma^2}{n}$$

where σ^2 is the variance between group means and n is the number of groups sampled. Alternatively, if a fraction f of the total number of groups is sampled, this may be replaced by

$$\frac{(1-f)\sigma^2}{n}$$

If there is no variation between the groups the variance of the estimated mean is then

$$\frac{\text{Variance within groups}}{\text{Number of samples taken}} = \frac{\sigma'^2}{nn'}$$

where σ'^2 is the variance within groups, and n' is the number of samples taken from each group. Again, if f' represents the fraction of each group sampled, this becomes

$$\frac{(1-f')\,\sigma'^2}{nn'}$$

If, now, there is variation both between and within the groups, both of these contribute to the variance of the mean. Since, however, the variation within groups will be reflected in the variation between groups, this latter portion only contributes, in part, to the variance of the estimated mean, which is

$$\frac{(1-f)\,\sigma^2}{n}+f\times\frac{(1-f')\,\sigma'^2}{nn'}$$

It should be noted that if f is sufficiently small to be neglected this becomes σ^2/n *i.e.* the same result as would be obtained using the group means as the sample and ignoring the variation within groups.

The analysis of a 2-stage sampling process may easily be carried out using a within and between group form of analysis. If the analysis of variance testing the difference between groups has a between group mean square of s^2 and a within group mean square of s'^2, then s^2/n' estimates σ^2 and s'^2 estimates σ'^2. Thus the estimated variance of the mean is

$$\frac{(1-f)\,s^2}{nn'}+\frac{f\,(1-f')\,s'^2}{nn'}$$

Sampling may similarly be carried out in several stages and corresponding formulae to the above exist for estimating the standard errors of estimated means. Such sampling is known as multi-stage sampling.

9A.10 *Ratio method in sampling*—It is often possible to make use of supplementary information in sampling to ensure that the sample is representative in certain observations. Thus, for example, if a random sample of individuals is taken from a population in which the proportion of males is known, the sample may be stratified for sex and, consequently, any variation in the estimated mean due to sex differences removed.

Frequently, stratification of a sample may be carried out with regard to supplementary observations, but sometimes this is not possible or other methods of utilizing the supplementary information are preferable. For instance, in estimating the number of people with a given attribute in the whole population it will be easier to estimate the proportion of people with the attribute in a given sample and to multiply this by the total size of population, if this is known. Here the supplementary information is the total size of population. As a second illustration, if it were required to estimate the total annual milk production of an area, it would, in fact, be easier to estimate the average milk production per cow and multiply this by the total number of cows, if this were known, than to try to estimate it directly from the sampled farms. Thus here the estimate of the total milk production would be

$$\frac{\text{Total milk production of sampled farms}}{\text{Number of cows on sampled farms}} \times \text{Total number of animals in population}$$

In general, we may derive an estimate of a sampled quantity from the formula

$$\frac{\text{Total of sampled quantity}}{\text{Total of concomitant variable in sample}} \times \text{Total of concomitant variable in population}$$

The use in this manner of supplementary information is called the ratio method.

There is a slight difficulty in attaching a standard error to an estimate obtained by the ratio method, since we have, in effect, to attach a standard error to the estimated ratio

$$\frac{\text{Total of sampled quantity}}{\text{Total of concomitant variable in sample}}$$

In the first of the above examples, where this ratio is

$$\frac{\text{No. of people with attribute in sample}}{\text{No. of people sampled}}$$

this presents no difficulty since the denominator is fixed in advance (unless, of course, a sample of households is being taken when the following remarks will apply). However, in the second of the above examples the ratio is

$$\frac{\text{Total milk production of sampled farms}}{\text{No. of cows on sampled farms}}$$

Here the denominator is not fixed in advance, and cows observed on the same farm cannot be regarded as independent observations since the treatment by the farmer of the animals may produce considerable improvement in milk production. To put this statement in alternative manner, the farm is the sampling unit and, consequently, possible variation in the denominator must be taken into account in estimating the standard error of the ratio.

In order to determine the variance of the ratio it is necessary to use an approximate formula which may be quoted as follows:

$$\text{Variance of } \frac{\bar{x}}{\bar{y}} = \frac{\bar{x}^2}{n\bar{y}^2}\left[\frac{\text{Variance of } x}{\bar{x}^2} + \frac{\text{Variance of } y}{\bar{y}^2} - \frac{2 \times \text{Covariance of } x \text{ and } y}{\bar{x}\,\bar{y}}\right]$$

For the above example, $\bar{x}$ is the mean milk production per farm, $\bar{y}$ is the mean number of cows per farm, $\bar{x}/\bar{y}$ is the estimated milk production per cow, n is the number of farms sampled. In order to calculate the standard error, it is necessary to estimate the variances of the milk production per farm and the number of animals per farm and also the covariance of the total milk production and number of animals per farm. These would then be substituted in the above formula.

To demonstrate the ratio method, suppose 30 farms were sampled with the following results:

Milk production x	10	60	6	43	74	193	0	43	87	.7
Number of cows y	3	15	2	10	17	41	0	12	21	2
Milk production x	16	2	0	153	61	0	0	29	63	41
Number of cows y	4	1	0	36	13	0	0	8	16	12
Milk production x	0	0	30	187	63	130	14	15	24	12
Number of cows y	0	0	8	42	16	28	6	4	8	3

$$\Sigma x = 1{,}363 \qquad \Sigma y = 328$$

$$\bar{x} = 45{\cdot}433 \qquad \bar{y} = 10{\cdot}933$$

$$\frac{\Sigma(x-\bar{x})^2}{29} = 3{,}021{\cdot}07 \qquad \frac{\Sigma(x-\bar{x})(y-\bar{y})}{29} = 662{\cdot}45 \qquad \frac{\Sigma(y-\bar{y})^2}{29} = 146{\cdot}55$$

$$\text{Mean milk production per cow} = 45{\cdot}433/10{\cdot}933 = 4{\cdot}156$$

$$\text{Standard error} = \sqrt{\left[\frac{(4{\cdot}156)^2}{30}\left\{\frac{3{,}021{\cdot}07}{(45{\cdot}433)^2} + \frac{146{\cdot}55}{(10{\cdot}933)^2} - \frac{2\,(662{\cdot}45)}{(45{\cdot}433)\,(10{\cdot}933)}\right\}\right]}$$

$$= \pm 0{\cdot}113$$

If the total number of cows in the district was known this could be used to multiply the mean milk production and its standard error, thus giving the estimated total milk production.

The use of ratios in this manner is not, of course, restricted to random sampling and it may be used for other methods. Again, where necessary, the factor $(1-f)$ may be used to adjust for the effect of finite populations. In the above example f would be the fraction of farms sampled.

9A.11 *Regression method in sampling*—Often when supplementary measurements are taken, if they are associated with the sampled measurements, a straight line relationship can be set up between the two sets of measurements. If the population mean of the supplementary measurements is known the corresponding population value for the sampled measurement may then be estimated. In this manner the estimate may be improved by 'standardizing' for the supplementary measurement. For instance, if we are sampling for a measurement which tends to increase linearly with age, we may estimate the form of association between age and the measurement. This may then be used to adjust the sample so that its mean age equals that of the population and so that it is, in effect, standardized for age.

This adjustment, which was in fact used in Chapter 7 as a method for improving the accuracy of comparisons between different groups, requires no new methods and may be made quite easily. It is particularly useful where two methods of taking measurements exist, one of which is expensive but accurate, the other being inexpensive but possibly biased. Both methods may then be carried out in a sample of the population and the cheaper method may be carried out over the whole population. The cheaper method may then be calibrated against the more exact method from the sample, and the value obtained for the whole population adjusted to give a more accurate estimate.

In particular, this approach may be used to calibrate eye estimates and to remove any bias in the estimate. This method has been called the regression method in sampling.

As an example we shall consider a series of eye estimates of the numbers of pine cones per tree e and the actual numbers of pine cones per tree n. On a random sample of 10 trees these gave the values:

Number of cones n	0	337	569	67	28	564	120	137	81	386
Estimated number of cones e	0	300	400	50	25	480	100	100	80	300

For these values, we get:

$$\Sigma n = 2{,}289 \qquad \Sigma e = 1.835$$

$$\bar{n} = 228{\cdot}9 \qquad \bar{e} = 183{\cdot}5$$

$$\Sigma(n-\bar{n})^2 = 425{,}473 \quad \Sigma(n-\bar{n})(e-\bar{e}) = 331{,}418 \quad \Sigma(e-\bar{e})^2 = 263{,}203$$

$$b = 331{,}418/263{,}203$$

$$= 1{\cdot}2592$$

and the regression equation of n on e is

$$n = 228{\cdot}9 + 1{\cdot}2592\,(e - 183{\cdot}5)$$

This may be tested by the analysis of variance:

	D.f.	*S.s.*	*M.s.*	*V.r.*
Regression	1	417,312	417,312	409
Residual	8	8,161	1,020·1	
Total	9	425,473		

There is obviously a bias in the eye estimates, which tend to be lower than the true numbers of cones. Consequently the above regression equation may be used to eliminate the bias from the mean eye estimate for the whole group of 184 trees of which this was a sample. The mean eye estimate for this group was 227·2 so that the corresponding number of cones is

$$228{\cdot}9 + 1{\cdot}2592\,(227{\cdot}2 - 183{\cdot}5) = 283{\cdot}9$$

Alternatively, this estimate may be regarded as the mean number of cones for the 10 sample trees corrected for the difference between the eye estimates for these trees and for the whole stand.

The standard error of this estimate may be derived in the usual manner from

$$\sqrt{\left[1{,}020{\cdot}1\left\{\frac{1}{10}+\frac{(227{\cdot}2-183{\cdot}5)^2}{263{,}203}\right\}\right]}=\pm 10{\cdot}5$$

The improvement in accuracy due to this method is evident when this standard error is compared with the standard error of the estimate obtained directly from the sample trees: $\pm 68{\cdot}8$.

It must however be noted that the above approach is subject to certain limitations. First, it is necessary for the association between the two sets of observations to be linear. Secondly, the population mean of the supplementary observation must be known. Thirdly, it assumes that the regression equation is derived from an unbiased sample of the whole population. Each of these limitations may be overcome (the first, only under certain conditions), but some modification of the form of analysis is required on each occasion.

SUMMARY OF PP 213 TO 218

Some of the sampling methods commonly used in surveys have been indicated. The appropriate formulae for estimating standard errors when the sampling is carried out in two or more stages have been derived.

Methods of using supplementary observations to improve estimates obtained by sampling have been given. In addition to stratification two other methods have been explained: the ratio method and the regression method. The ratio method effectively estimates an unkown ratio and the standard error of this estimate, and utilizes this in deriving an improved estimate. The regression method does the same with a regression equation. Examples have been given of the application of these methods.

EXAMPLES

100 In a sampling investigation to estimate the yield of wheat on a given area, the area was divided into 75 sub-areas which were stratified into 5 blocks of 15 sub-areas and 3 sub-areas were chosen at random from each block. These fifteen sub-areas were then sampled by taking two random samples of one tenth of their area. The following table of results was obtained:

Sub-area	*1*		*2*		*3*	
Block \ *Sample*	*a*	*b*	*a*	*b*	*a*	*b*
I	655	690	495	670	426	466
II	440	485	450	525	555	505
III	635	560	469	562	735	665
IV	600	550	685	655	695	695
V	533	587	505	430	765	555

Construct the analysis of variance testing the differences between blocks and sub-areas:

	D.f.	*S.s.*	*M.s.*	*V.r.*
Blocks	4	77,316	19,329	4·88
Residual between sub-areas	10	143,970	14,397	3·64
Between sub-areas	14	221,286	———	——
Within sub-areas	15	59,407	3,960	
Total	29	280,693		

Hence show that the mean yield is 574·8 and that its standard error is

$$\sqrt{\left[\frac{(1-1/5)\,14{,}397}{30}+\frac{1/5\,(1-1/5)\,3{,}960}{30}\right]}=\pm 20{\cdot}1$$

(The true value here was in fact 587·9.)

101 Show that if the ratio method is used on the cone data of section 9A.11, the estimated mean number of cones, ignoring the factor $(1-f)$, is $283{\cdot}4 \pm 11{\cdot}82$.

102 Show that if the regression method is used on the milk production data of section 9A.10, and the mean number of cows per farm is 12·3, the estimated mean milk production per farm is $51{\cdot}61 \pm 0{\cdot}964$.

BIBLIOGRAPHY

CHAPTER 1

For detailed accounts of the presentation of sets of measurements see

CROXTON, F. E. and COWDEN, D. J. *Applied General Statistics* pp 1-264 New York, 1939

CONNOR, L. R. *Statistics in Theory and Practice* pp 1-113 London, 1938

RICHARDSON, C. H. *An Introduction to Statistical Analysis* pp 1-173 New York, 1944

SIMPSON, G. G. and ROE, A. *Quantitative Zoology* New York, 1939

CHAPTER 2

For a more detailed account of applications of the normal distribution see

GARRETT, H. E. *Statistics in Psychology and Education* New York, 1937

For an interesting proof of the normal law from the principle of the arithmetic mean see

BRUNT, D. *The Combination of Observations* Cambridge, 1931

For derivation and mathematical investigation of binomial and Poisson distributions see

DAVID, F. N. *Probability Theory for Statistical Methods* Cambridge, 1949

CHAPTERS 3 AND 4

For testing differences between means and discussion of experimental design see

GOULDEN, C. H. *Methods of Statistical Analysis* New York, 1939

FISHER, R. A. *The Design of Experiments* Edinburgh, 1947

WISHART, J. Field Trials, their Lay-out and Statistical Analysis *Commonwealth Bureau of Plant Breeding and Genetics*

YATES, F. The Design and Analysis of Factorial Experiments *Commonwealth Bureau of Soil Science, Technical Communication No.* 35

CHAPTER 5

For analysis and application in genetics of the chi-squared test see

MATHER, K. *Statistical Analysis in Biology* London, 1943

FISHER, R. A. *Statistical Methods for Research Workers* Edinburgh, 1941

CHAPTER 6

For detailed accounts of regression and correlation theory see

TIPPETT, L. H. C. *The Methods of Statistics* London, 1937

KENNEY, J. F. *The Mathematics of Statistics* New York, 1939

For correlation and partial correlation coefficients see

YULE, G. U. and KENDALL, M. G. *An Introduction to the Theory of Statistics* London, 1946

For particular applications of the method of regression see

LYLE, P. *Regression Analysis of Production Costs and Factory Operations* Edinburgh, 1944

CHAPTER 7

For methods of the analysis of variance and covariance see

SNEDECOR, G. W. *Statistical Methods* Iowa, 1946

This book deals at length with the analysis of non-orthogonal data and with the testing of individual degrees of freedom.

CHAPTER 8

For specialized methods of dealing with ranks see

KENDALL, M. G. *Rank Correlation Methods* London, 1948

Similar accounts of methods for dealing with percentages are provided by

FINNEY, D. J. *Probit Analysis* Cambridge, 1947

EMMENS, C. W. *Principles of Biological Assay* London, 1948

Various types of non-normal frequency curves are classified in

ELDERTON, W. P. *Frequency Curves and Correlation* Cambridge, 1938

For an account of, and a list of references to, work on non-normality see

GEARY, R. C. *Biometrika* 34 (1947) 209-242

CHAPTER 9

For an up to date account of sampling methods see

YATES, F. *Sampling Methods for Censuses and Surveys* New York, 1949

For application of these methods to forestry see

SCHUMACHER, F. X. and CHAPMAN, R. A. *Sampling Methods in Forestry and Range Management* Durham, N. C., 1942

For a description of the problems in survey work see

The Preparation of Sampling Survey Reports *Statistical Papers Series C, No.* 1 Statistical Office of the United Nations

TABLES

FISHER, R. A. and YATES, F. *Statistical Tables for Biological, Agricultural and Medical Research* London, 1943

To find the existing tables of different functions see

FLETCHER, A., MILLER, J. C. P. and ROSENHEAD, L. *An Index of Mathematical Tables* London, 1946

For general purposes see

BARLOW, P. *Tables of Squares, Cubes etc.* London, 1930

MILNE-THOMSON, L. M. and COMRIE, L. J. *Standard Four-Figure Mathematical Tables* London, 1931

For individual tables of statistical functions see

Tracts for Computers Nos. 1-25 Department of Statistics, University College, London

Biometrika Series of Statistical Tables, published by the Editors of *Biometrika*, University College, London

APPENDIX

Table I

Table of the Percentage of Observations exceeding a Given Normal deviate d

d	*Percentage*	*d*	*Percentage*	*d*	*Percentage*
4·0	0·003	1·0	15·9	−1·1	86·43
3·5	0·023	0·9	18·4	−1·2	88·49
3·2	0·069	0·8	21·2	−1·3	90·32
3·0	0·135	0·7	24·2	−1·4	91·92
2·8	0·256	0·6	27·4	−1·5	93·32
2·6	0·466	0·5	30·9	−1·6	94·52
2·5	0·621	0·4	34·5	−1·7	95·54
2·4	0·820	0·3	38·2	−1·8	96·41
2·3	1·07	0·2	42·1	−1·9	97·13
2·2	1·39	0·1	46·0	−2·0	97·72
2·1	1·79	0·0	50·0	−2·1	98·21
2·0	2·28	−0·1	54·0	−2·2	98·61
1·9	2·87	−0·2	57·9	−2·3	98·93
1·8	3·59	−0·3	61·8	−2·4	99·180
1·7	4·46	−0·4	65·5	−2·5	99·379
1·6	5·48	−0·5	69·1	−2·6	99·534
1·5	6·68	−0·6	72·6	−2·8	99·744
1·4	8·08	−0·7	75·8	−3·0	99·865
1·3	9·68	−0·8	78·8	−3·2	99·931
1·2	11·51	−0·9	81·6	−3·5	99·977
1·1	13·57	−1·0	84·1	−4·0	99·997

Table II

Table of Limits for the Deviate d corresponding to a Given Percentage

Percentage	±*d*	*Percentage*	±*d*	*Percentage*	±*d*
99·99	3·89	85	1·44	40	0·52
99·9	3·29	80	1·28	35	0·45
99	2·58	75	1·15	30	0·39
98	2·33	70	1·04	25	0·32
97	2·17	65	0·93	20	0·25
96	2·05	60	0·84	15	0·19
95	1·96	55	0·76	10	0·13
92	1·75	50	0·67	5	0·06
90	1·64	45	0·60	0	0·00

Table III

Variance-Ratio Table, 5 per cent Points giving the Values of the Ratio exceeded by Pure Chance in 5 per cent of Trials

Degrees of freedom of denominator	Degrees of freedom of numerator									
	1	2	3	4	5	6	7	8	9	10
1	161	200	216	225	230	234	237	239	241	242
2	18·51	19·00	19·16	19·25	19·30	19·33	19·35	19·37	19·38	19·40
3	10·13	9·55	9·28	9·12	9·01	8·94	8·89	8·84	8·81	8·78
4	7·71	6·94	6·59	6·39	6·26	6·16	6·09	6·04	6·00	5·96
5	6·61	5·79	5·41	5·19	5·05	4·95	4·88	4·82	4·77	4·74
6	5·99	5·14	4·76	4·53	4·39	4·28	4·21	4·15	4·10	4·00
7	5·59	4·74	4·35	4·12	3·97	3·87	3·79	3·73	3·68	3·64
8	5·32	4·46	4·07	3·84	3·69	3·58	3·50	3·44	3·39	3·35
9	5·12	4·26	3·86	3·63	3·48	3·37	3·29	3·23	3·18	3·14
10	4·96	4·10	3·71	3·48	3·33	3·22	3·14	3·07	3·02	2·98
11	4·84	3·98	3·59	3·36	3·20	3·09	3·01	2·95	2·90	2·85
12	4·75	3·88	3·49	3·26	3·11	3·00	2·91	2·85	2·80	2·75
13	4·67	3·80	3·41	3·18	3·02	2·92	2·83	2·77	2·71	2·67
14	4·60	3·74	3·34	3·11	2·96	2·85	2·76	2·70	2·65	2·60
15	4·54	3·68	3·29	3·06	2·90	2·79	2·71	2·64	2·59	2·54
16	4·49	3·63	3·24	3·01	2·85	2·74	2·66	2·59	2·54	2·49
17	4·45	3·59	3·20	2·96	2·81	2·70	2·61	2·55	2·49	2·45
18	4·41	3·55	3·16	2·93	2·77	2·66	2·58	2·51	2·46	2·41
19	4·38	3·52	3·13	2·90	2·74	2·63	2·54	2·48	2·42	2·38
20	4·35	3·49	3·10	2·87	2·71	2·60	2·51	2·45	2·39	2·35
21	4·32	3·47	3·07	2·84	2·68	2·57	2·49	2·42	2·37	2·32
22	4·30	3·44	3·05	2·82	2·66	2·55	2·46	2·40	2·34	2·30
23	4·28	3·42	3·03	2·80	2·64	2·53	2·44	2·37	2·32	2·27
24	4·26	3·40	3·01	2·78	2·62	2·51	2·42	2·36	2·30	2·25
25	4·24	3·38	2·99	2·76	2·60	2·49	2·40	2·34	2·28	2·24
26	4·22	3·37	2·98	2·74	2·59	2·47	2·39	2·32	2·27	2·22
27	4·21	3·35	2·96	2·73	2·57	2·46	2·37	2·30	2·25	2·20
28	4·20	3·34	2·95	2·71	2·56	2·44	2·36	2·29	2·24	2·19
29	4·18	3·33	2·93	2·70	2·54	2·43	2·35	2·28	2·22	2·18
30	4·17	3·32	2·92	2·69	2·53	2·42	2·33	2·27	2·21	2·16
40	4·08	3·23	2·84	2·61	2·45	2·34	2·25	2·18	2·12	2·08
50	4·03	3·18	2·79	2·56	2·40	2·29	2·20	2·13	2·07	2·02
60	4·00	3·15	2·76	2·52	2·37	2·25	2·17	2·10	2·04	1·99
70	3·98	3·13	2·74	2·50	2·35	2·23	2·14	2·07	2·01	1·97
80	3·96	3·11	2·72	2·48	2·33	2·21	2·12	2·05	1·99	1·95
100	3·94	3·09	2·70	2·46	2·30	2·19	2·10	2·03	1·97	1·92
150	3·91	3·06	2·67	2·43	2·27	2·16	2·07	2·00	1·94	1·89
200	3·89	3·04	2·65	2·41	2·26	2·14	2·05	1·98	1·92	1·87
∞	3·84	3·00	2·60	2·37	2·21	2·09	2·01	1·94	1·88	1·83

Table III (continued)

Variance-Ratio Table, 5 per cent Points giving the Values of the Ratio exceeded by Pure Chance in 5 per cent of Trials

Degrees of freedom of numerator										
12	16	20	24	30	40	50	60	75	100	∞
244	246	248	249	250	251	252	252	253	253	254
19·41	19·43	19·45	19·45	19·46	19·47	19·47	19·48	19·48	19·49	19·50
8·74	8·69	8·66	8·64	8·62	8·59	8·58	8·57	8·57	8·56	8·53
5·91	5·84	5·80	5·77	5·74	5·72	5·70	5·69	5·68	5·66	5·63
4·68	4·60	4·56	4·53	4·50	4·46	4·44	4·43	4·42	4·40	4·36
4·00	3·92	3·87	3·84	3·81	3·77	3·75	3·74	3·72	3·71	3·67
3·57	3·49	3·44	3·41	3·38	3·34	3·32	3·30	3·29	3·28	3·23
3·28	3·20	3·15	3·12	3·08	3·04	3·03	3·00	3·00	2·98	2·93
3·07	2·98	2·94	2·90	2·86	2·82	2·80	2·79	2·77	2·76	2·71
2·91	2·82	2·77	2·74	2·70	2·66	2·64	2·62	2·61	2·59	2·54
2·79	2·70	2·65	2·61	2·57	2·53	2·50	2·49	2·47	2·45	2·40
2·69	2·60	2·54	2·50	2·47	2·42	2·40	2·38	2·36	2·35	2·30
2·60	2·51	2·46	2·42	2·38	2·34	2·32	2·30	2·28	2·26	2·21
2·53	2·44	2·39	2·35	2·31	2·27	2·24	2·22	2·21	2·19	2·13
2·48	2·39	2·33	2·29	2·25	2·20	2·18	2·16	2·15	2·12	2·07
2·42	2·33	2·28	2·24	2·19	2·15	2·13	2·10	2·09	2·07	2·01
2·38	2·29	2·23	2·19	2·15	2·10	2·08	2·06	2·04	2·02	1·96
2·34	2·25	2·19	2·15	2·11	2·06	2·04	2·02	2·00	1·98	1·92
2·31	2·21	2·16	2·11	2·07	2·03	2·00	1·98	1·96	1·94	1·88
2·28	2·18	2·12	2·08	2·04	1·99	1·96	1·95	1·92	1·90	1·84
2·25	2·15	2·10	2·05	2·01	1·96	1·93	1·92	1·89	1·87	1·81
2·23	2·13	2·07	2·03	1·98	1·94	1·91	1·89	1·87	1·84	1·78
2·20	2·10	2·05	2·06	1·96	1·91	1·88	1·86	1·84	1·82	1·76
2·18	2·09	2·03	1·98	1·94	1·89	1·86	1·84	1·82	1·80	1·73
2·16	2·06	2·01	1·96	1·92	1·87	1·84	1·82	1·80	1·77	1·71
2·15	2·05	1·99	1·95	1·90	1·85	1·82	1·80	1·78	1·76	1·69
2·13	2·03	1·97	1·93	1·88	1·84	1·80	1·78	1·76	1·74	1·67
2·12	2·02	1·96	1·91	1·87	1·82	1·78	1·77	1·75	1·72	1·65
2·10	2·00	1·94	1·90	1·85	1·80	1·77	1·75	1·73	1·71	1·64
2·09	1·99	1·93	1·89	1·84	1·79	1·76	1·74	1·72	1·69	1·62
2·00	1·90	1·84	1·79	1·74	1·69	1·66	1·64	1·61	1·59	1·51
1·95	1·85	1·78	1·74	1·69	1·63	1·60	1·58	1·55	1·52	1·44
1·92	1·81	1·75	1·70	1·65	1·59	1·56	1·53	1·50	1·48	1·39
1·89	1·79	1·72	1·67	1·62	1·56	1·53	1·50	1·47	1·45	1·35
1·88	1·77	1·70	1·65	1·60	1·54	1·51	1·48	1·45	1·42	1·32
1·85	1·75	1·68	1·63	1·57	1·51	1·48	1·45	1·42	1·39	1·28
1·82	1·71	1·64	1·59	1·54	1·47	1·44	1·41	1·37	1·34	1·22
1·80	1·69	1·62	1·57	1·52	1·45	1·42	1·39	1·35	1·32	1·19
1·75	1·64	1·57	1·52	1·46	1·39	1·35	1·32	1·28	1·24	1·00

Table IV

Variance-Ratio Table, 1 per cent Points giving the Values of the Ratio exceeded by Pure Chance in 1 per cent of Trials

Degrees of freedom of denominator	Degrees of freedom of numerator 1	2	3	4	5	6	7	8	9	10
1	4,052	4,999	5,403	5,625	5,764	5,859	5,928	5,982	6,022	6,056
2	98·50	99·00	99·17	99·25	99·30	99·33	99·36	99·36	99·39	99·40
3	34·12	30·82	29·46	28·71	28·24	27·91	27·67	27·49	27·34	27·23
4	21·20	18·00	16·69	15·98	15·52	15·21	14·98	14·80	14·66	14·55
5	16·26	13·27	12·06	11·39	10·97	10·67	10·46	10·29	10·16	10·05
6	13·74	10·92	9·78	9·15	8·75	8·47	8·26	8·10	7·98	7·87
7	12·25	9·55	8·45	7·85	7·46	7·19	6·99	6·84	6·72	6·62
8	11·26	8·65	7·59	7·01	6·63	6·37	6·18	6·03	5·91	5·81
9	10·56	8·02	6·99	6·42	6·06	5·80	5·61	5·47	5·35	5·26
10	10·04	7·56	6·55	5·99	5·64	5·39	5·20	5·06	4·94	4·85
11	9·65	7·20	6·22	5·67	5·32	5·07	4·89	4·74	4·63	4·54
12	9·33	6·93	5·95	5·41	5·06	4·82	4·64	4·50	4·39	4·30
13	9·07	6·70	5·74	5·20	4·86	4·62	4·44	4·30	4·19	4·10
14	8·86	6·51	5·56	5·04	4·69	4·46	4·28	4·14	4·03	3·94
15	8·68	6·36	5·42	4·89	4·56	4·32	4·14	4·00	3·89	3·80
16	8·53	6·23	5·29	4·77	4·44	4·20	4·03	3·89	3·78	3·69
17	8·40	6·11	5·18	4·67	4·34	4·10	3·93	3·79	3·68	3·59
18	8·28	6·01	5·09	4·58	4·25	4·01	3·84	3·71	3·60	3·51
19	8·18	5·93	5·01	4·50	4·17	3·94	3·76	3·63	3·52	3·43
20	8·10	5·85	4·94	4·43	4·10	3·87	3·70	3·56	3·46	3·37
21	8·02	5·78	4·87	4·37	4·04	3·81	3·64	3·51	3·40	3·31
22	7·94	5·72	4·82	4·31	3·99	3·76	3·59	3·45	3·36	3·26
23	7·88	5·66	4·76	4·26	3·94	3·71	3·54	3·41	3·30	3·21
24	7·82	5·61	4·72	4·22	3·90	3·67	3·50	3·36	3·26	3·17
25	7·77	5·57	4·68	4·18	3·86	3·63	3·46	3·32	3·22	3·13
26	7·72	5·53	4·64	4·14	3·82	3·59	3·42	3·29	3·18	3·09
27	7·68	5·49	4·60	4·11	3·78	3·56	3·39	3·26	3·15	3·06
28	7·64	5·45	4·57	4·07	3·75	3·53	3·36	3·23	3·12	3·03
29	7·60	5·42	4·54	4·04	3·73	3·50	3·33	3·20	3·09	3·00
30	7·56	5·39	4·51	4·02	3·70	3·47	3·30	3·17	3·07	2·98
40	7·31	5·18	4·31	3·83	3·51	3·29	3·12	2·99	2·89	2·80
50	7·17	5·06	4·20	3·72	3·41	3·18	3·02	2·88	2·78	2·70
60	7·08	4·98	4·13	3·65	3·34	3·12	2·95	2·82	2·72	2·63
70	7·01	4·92	4·08	3·60	3·29	3·07	2·91	2·77	2·67	2·59
80	6·96	4·88	4·04	3·56	3·25	3·04	2·87	2·74	2·64	2·55
100	6·90	4·82	3·98	3·51	3·20	2·99	2·82	2·69	2·59	2·51
150	6·81	4·75	3·91	3·44	3·14	2·92	2·76	2·62	2·53	2·44
200	6·76	4·71	3·88	3·41	3·11	2·90	2·73	2·60	2·50	2·41
∞	6·63	4·60	3·78	3·32	3·02	2·80	2·64	2·51	2·41	2·32

Table IV (continued)

Variance-Ratio Table, 1 per cent Points giving the Values of the Ratio exceeded by Pure Chance in 1 per cent of Trials

Degrees of freedom of numerator										
12	16	20	24	30	40	50	60	75	100	∞
6,106	6,169	6,209	6,235	6,261	6,287	6,302	6,313	6,323	6,334	6,366
99·42	99·44	99·45	99·46	99·47	99·47	99·48	99·48	99·49	99·49	99·50
27·05	26·83	26·69	26·60	26·50	26·41	26·35	26·32	26·27	26·23	26·12
14·37	14·15	14·02	13·93	13·84	13·74	13·69	13·65	13·61	13·57	13·46
9·89	9·68	9·55	9·47	9·38	9·29	9·24	9·20	9·17	9·13	9·02
7·72	7·52	7·40	7·31	7·23	7·14	7·09	7·06	7·02	6·99	6·88
6·47	6·27	6·16	6·07	5·99	5·91	5·85	5·82	5·78	5·75	5·65
5·67	5·48	5·36	5·28	5·20	5·12	5·06	5·03	5·00	4·96	4·86
5·11	4·92	4·81	4·73	4·65	4·57	4·51	4·48	4·45	4·41	4·31
4·71	4·52	4·40	4·33	4·25	4·16	4·12	4·08	4·05	4·01	3·91
4·40	4·21	4·10	4·02	3·94	3·86	3·80	3·78	3·74	3·70	3·60
4·16	3·98	3·86	3·78	3·70	3·62	3·56	3·54	3·49	3·46	3·36
3·96	3·78	3·66	3·59	3·51	3·42	3·37	3·34	3·30	3·27	3·16
3·80	3·62	3·50	3·43	3·35	3·27	3·21	3·18	3·14	3·11	3·00
3·67	3·48	3·37	3·29	3·21	3·13	3·07	3·05	3·00	2·97	2·87
3·55	3·37	3·26	3·18	3·10	3·02	2·96	2·93	2·89	2·86	2·75
3·45	3·27	3·16	3·08	3·00	2·92	2·86	2·83	2·79	2·76	2·65
3·37	3·19	3·08	3·00	2·91	2·84	2·78	2·75	2·71	2·68	2·57
3·30	3·12	3·00	2·92	2·84	2·76	2·70	2·67	2·63	2·60	2·49
3·23	3·05	2·94	2·86	2·78	2·69	2·63	2·61	2·56	2·53	2·42
3·17	2·99	2·88	2·80	2·72	2·64	2·58	2·55	2·51	2·47	2·36
3·12	2·94	2·83	2·75	2·67	2·58	2·53	2·50	2·46	2·42	2·31
3·07	2·89	2·78	2·70	2·62	2·54	2·48	2·45	2·41	2·37	2·26
3·03	2·85	2·74	2·66	2·58	2·49	2·44	2·40	2·36	2·33	2·21
2·99	2·81	2·70	2·62	2·54	2·45	2·40	2·36	2·32	2·29	2·17
2·96	2·77	2·66	2·58	2·50	2·42	2·36	2·33	2·28	2·25	2·13
2·93	2·74	2·63	2·55	2·47	2·38	2·33	2·29	2·25	2·21	2·10
2·90	2·71	2·60	2·52	2·44	2·35	2·30	2·26	2·22	2·18	2·06
2·87	2·68	2·57	2·49	2·41	2·32	2·27	2·23	2·19	2·15	2·03
2·84	2·66	2·55	2·47	2·39	2·30	2·24	2·21	2·16	2·13	2·01
2·66	2·49	2·37	2·29	2·20	2·11	2·05	2·02	1·97	1·94	1·80
2·56	2·39	2·26	2·18	2·10	2·00	1·94	1·91	1·86	1·82	1·68
2·50	2·32	2·20	2·12	2·03	1·94	1·87	1·84	1·79	1·74	1·60
2·45	2·28	2·15	2·07	1·98	1·88	1·82	1·79	1·74	1·69	1·53
2·41	2·24	2·11	2·03	1·94	1·84	1·78	1·74	1·70	1·65	1·49
2·36	2·19	2·06	1·98	1·89	1·79	1·73	1·69	1·64	1·59	1·43
2·30	2·12	2·00	1·91	1·83	1·72	1·66	1·62	1·56	1·51	1·33
2·28	2·09	1·97	1·88	1·79	1·69	1·62	1·58	1·53	1·48	1·28
2·18	1·99	1·88	1·79	1·70	1·59	1·52	1·47	1·41	1·36	1·00

Table V

Percentage of Trials in which a Given Estimated Deviate, t, is Exceeded

Degrees of freedom	*Percentage of trials in which deviate is exceeded**							
	50	25	10	5	2·5	1·0	0·5	0·1
1	1·00	2·41	6·31	12·71	25·45	63·66	127·32	636·62
2	0·82	1·60	2·92	4·30	6·20	9·92	14·09	31·60
3	0·76	1·42	2·35	3·18	4·18	5·84	7·45	12·94
4	0·74	1·34	2·13	2·78	3·50	4·60	5·60	8·61
5	0·73	1·30	2·01	2·57	3·16	4·03	4·77	6·86
6	0·72	1·27	1·94	2·45	2·97	3·71	4·32	5·96
7	0·71	1·25	1·89	2·36	2·84	3·50	4·03	5·40
8	0·71	1·24	1·86	2·30	2·75	3·35	3·83	5·04
9	0·70	1·23	1·83	2·26	2·68	3·25	3·69	4·78
10	0·70	1·22	1·81	2·23	2·63	3·17	3·58	4·59
11	0·70	1·21	1·80	2·20	2·59	3·11	3·50	4·44
12	0·70	1·21	1·78	2·18	2·56	3·05	3·43	4·32
13	0·69	1·20	1·77	2·16	2·53	3·01	3·37	4·22
14	0·69	1·20	1·76	2·14	2·51	2·98	3·32	4·14
15	0·69	1·20	1·75	2·13	2·49	2·95	3·29	4·07
16	0·69	1·19	1·74	2·12	2·47	2·92	3·25	4·01
17	0·69	1·19	1·74	2·11	2·46	2·90	3·22	3·96
18	0·69	1·19	1·73	2·10	2·44	2·88	3·20	3·92
19	0·69	1·19	1·73	2·09	2·43	2·86	3·17	3·88
20	0·69	1·18	1·72	2·09	2·42	2·84	3·15	3·85
22	0·69	1·18	1·72	2·07	2·40	2·82	3·12	3·79
24	0·68	1·18	1·71	2·06	2·39	2·80	3·09	3·75
26	0·68	1·17	1·71	2·06	2·38	2·78	3·07	3·71
28	0·68	1·17	1·70	2·05	2·37	2·76	3·05	3·67
30	0·68	1·17	1·70	2·04	2·36	2·75	3·03	3·65
40	0·68	1·17	1·68	2·02	2·33	2·70	2·97	3·55
50	0·68	1·16	1·68	2·01	2·31	2·68	2·93	3·50
60	0·68	1·16	1·67	2·00	2·30	2·66	2·91	3·46
∞	0·67	1·15	1·64	1·96	2·24	2·58	2·81	3·29

*For high percentages the normal deviate provides a good approximation. Alternatively use the approximation of section 8A.8.

Table VI
Table of χ^2 Distribution

Degrees of freedom n	*Percentage of trials in which χ^2 is exceeded*							
	50	25	10	5	2·5	1·0	0·5	0·1
1	0·45	1·32	2·70	3·84	5·02	6·63	7·88	10·83
2	1·39	2·77	4·60	5·99	7·38	9·21	10·60	13·82
3	2·36	4·11	6·25	7·81	9·35	11·34	12·84	16·27
4	3·36	5·38	7·78	9·49	11·14	13·28	14·86	18·46
5	4·35	6·62	9·24	11·07	12·83	15·09	16·75	20·52
6	5·35	7·84	10·64	12·59	14·45	16·81	18·55	22·46
7	6·34	9·04	12·02	14·07	16·01	18·48	20·28	24·32
8	7·34	10·22	13·36	15·51	17·53	20·09	21·96	26·12
9	8·34	11·39	14·68	16·92	19·02	21·67	23·59	27·88
10	9·34	12·55	15·99	18·31	20·48	23·21	25·19	29·59
11	10·34	13·70	17·28	19·68	21·92	24·72	26·76	31·26
12	11·34	14·84	18·55	21·03	23·34	26·22	28·30	32·91
13	12·34	15·98	19·81	22·36	24·74	27·69	29·82	34·53
14	13·34	17·12	21·06	23·68	26·12	29·14	31·32	36·12
15	14·34	18·24	22·31	25·00	27·49	30·58	32·80	37·70
16	15·34	19·37	23·54	26·30	28·84	32·00	34·27	39·25
17	16·34	20·49	24·77	27·59	30·19	33·41	35·72	40·79
18	17·34	21·60	25·99	28·87	31·53	34·80	37·16	42·31
19	18·34	22·72	27·20	30·14	32·85	36·19	38·58	43·82
20	19·34	23·83	28·41	31·41	34·17	37·57	40·00	45·32
22	21·34	26·04	30·81	33·92	36·78	40·29	42·80	48·27
24	23·34	28·24	33·20	36·42	39·36	42·98	45·56	51·18
26	25·34	30·43	35·56	38·88	41·92	45·64	48·29	54·05
28	27·34	32·62	37·92	41·34	44·46	48·28	50·99	56·89
30*	29·34	34·80	40·26	43·77	46·98	50·89	53·67	59·70

Example

*For more than 30 degrees of freedom use the fact that $\sqrt{(2\chi^2)} - \sqrt{(2n-1)}$ is approximately a normal deviate.

A χ^2 of 45 degrees of freedom takes a value 74·23. Then $\sqrt{(2 \times 74{\cdot}23)} - \sqrt{(2 \times 45 - 1)} = 2{\cdot}75$ is approximately a normal deviate. Using *Table II* it is seen that as high a value as this occurs in less than 1 per cent of trials by pure chance.

Table VII
Table of Random Numbers

3 4 8 4 8	1 6 6 4 6	2 8 3 4 9	7 0 0 8 0	1 9 8 2 3
7 3 9 3 8	4 5 3 1 6	2 4 8 4 3	0 1 1 2 9	5 6 7 9 5
9 8 6 4 7	8 4 7 2 6	2 1 4 4 8	1 2 6 1 4	5 9 8 5 5
0 2 7 6 8	7 9 9 8 8	9 2 1 6 3	5 5 1 5 2	2 8 0 2 9
1 0 5 8 9	7 1 3 4 9	4 8 6 4 9	4 2 3 6 9	6 3 3 5 0
0 2 2 3 6	1 5 6 8 4	7 5 4 5 5	9 2 4 6 9	6 4 1 4 7
4 7 2 0 0	0 5 3 2 8	9 2 5 8 2	9 2 1 3 4	1 2 3 0 3
3 9 0 9 9	1 3 0 4 1	2 0 7 3 8	4 6 1 2 0	9 7 0 2 8
4 5 8 9 8	6 7 6 2 2	3 1 2 4 6	4 7 7 9 7	6 9 5 9 4
7 8 3 5 5	8 6 9 3 3	1 1 2 1 5	0 5 7 3 7	3 9 1 4 3
4 2 0 1 7	6 7 3 8 1	3 5 6 9 0	8 6 0 9 7	5 1 8 5 8
1 5 8 0 3	4 5 9 5 1	6 3 9 5 4	3 3 9 7 2	4 4 0 3 4
9 4 7 6 8	8 2 0 3 7	9 4 5 2 6	0 4 1 9 1	8 7 2 2 8
9 7 4 2 9	0 1 5 5 6	2 0 5 8 5	3 4 6 3 5	5 3 4 0 2
4 4 1 5 8	8 6 3 6 0	0 4 4 8 6	8 2 3 4 3	1 6 6 9 8
0 2 9 7 4	0 3 9 9 0	4 8 4 1 2	6 1 7 6 5	5 2 2 2 8
9 7 8 0 8	6 7 6 0 1	7 0 3 5 5	0 0 1 8 2	3 4 8 1 9
7 8 1 7 7	9 3 8 5 2	4 2 9 7 2	0 1 8 9 3	9 7 6 1 0
1 8 0 4 3	9 0 5 6 1	7 3 2 9 2	4 0 3 5 0	3 9 2 6 5
4 0 4 6 2	8 7 6 1 2	9 2 3 8 1	7 0 2 3 4	2 2 4 8 3
9 8 1 9 4	3 6 2 3 2	5 1 9 2 2	6 4 8 3 5	1 0 4 1 4
9 7 5 3 5	0 0 0 4 8	7 6 3 0 9	9 0 1 4 4	9 3 3 5 1
7 9 7 6 7	9 0 7 7 1	4 1 0 8 7	7 0 6 0 0	5 2 3 8 6
2 1 4 4 1	0 1 2 9 2	5 9 9 4 0	2 6 3 3 1	6 3 3 7 3
7 7 4 3 8	1 6 9 0 6	3 2 9 6 1	8 7 4 0 4	5 6 8 8 0
2 9 1 3 6	0 7 8 6 0	8 4 3 1 7	6 3 1 4 7	1 7 8 0 1
0 9 6 7 7	6 6 4 7 7	7 1 2 8 8	2 1 6 8 6	2 5 0 3 3
8 4 4 0 3	6 0 7 3 4	5 7 5 4 2	7 1 6 0 1	2 1 4 4 8
8 0 8 5 1	0 2 5 1 2	9 4 7 9 3	6 7 6 2 0	4 4 0 4 5
2 7 7 0 3	1 9 3 6 2	4 4 0 2 9	3 0 8 3 9	5 4 2 6 7
2 1 9 4 5	7 5 0 2 3	3 2 8 2 4	4 5 0 5 4	5 4 0 5 0
3 5 3 0 7	9 3 8 6 0	7 1 5 6 1	5 8 5 0 6	7 3 9 0 8
9 7 8 5 6	9 3 0 8 3	3 7 9 4 7	5 0 5 3 9	3 1 6 9 8
3 9 4 4 8	9 5 8 9 9	8 9 9 4 6	1 5 1 9 4	2 0 4 2 6
7 0 0 6 7	0 9 5 1 4	0 5 8 2 2	0 4 4 9 2	6 8 6 2 1
8 7 7 2 5	6 1 4 6 2	3 4 4 0 5	6 7 5 9 1	2 5 5 6 3
5 1 7 6 7	5 6 4 2 5	4 3 0 7 0	1 0 5 6 5	1 9 9 8 2
2 7 4 3 8	1 1 6 7 4	3 3 6 3 1	4 9 2 6 4	1 2 8 6 2
3 8 3 9 2	4 5 8 7 8	0 2 0 7 0	5 4 1 0 6	3 8 4 6 2
3 4 5 2 2	8 3 4 7 5	4 6 4 3 8	7 6 9 3 2	2 1 5 8 3

Table VII (continued)
Table of Random Numbers

5 8 3 4 1	1 7 0 1 6	2 1 3 7 2	0 8 9 9 4	1 1 0 6 4
1 2 2 9 8	5 2 5 9 5	6 6 5 5 8	4 2 4 3 7	9 0 9 1 3
9 4 5 4 4	1 7 0 9 2	9 4 1 0 4	0 5 1 7 4	7 6 9 4 3
5 2 1 2 0	8 8 6 3 0	9 8 6 6 5	8 4 4 9 2	3 5 5 9 6
5 8 0 8 5	2 6 3 4 1	0 9 3 6 8	3 2 8 8 1	7 4 8 8 0
2 4 1 3 5	9 9 7 0 4	5 1 5 4 2	1 1 0 5 2	7 9 9 5 7
2 2 1 3 1	6 5 0 3 1	2 6 5 0 5	8 0 5 6 9	3 2 3 3 8
0 7 7 4 5	0 0 5 3 3	4 2 3 4 5	0 4 8 1 1	0 3 7 7 8
4 4 6 0 9	8 7 7 7 9	2 4 8 6 3	7 8 9 4 0	6 5 0 5 6
9 9 4 8 4	2 9 6 0 4	0 2 6 3 6	7 3 1 8 4	0 0 8 3 8
5 1 0 0 6	2 8 1 8 3	9 8 2 5 9	0 5 1 7 8	3 7 5 5 0
1 6 1 4 4	7 6 0 3 1	6 6 7 0 3	4 7 5 7 5	5 2 2 5 6
1 5 3 5 0	2 5 7 6 4	2 3 7 9 0	7 5 0 4 8	6 7 4 0 5
0 5 7 5 5	9 7 0 9 3	2 0 7 4 7	1 9 2 2 1	3 4 2 4 6
4 0 5 9 2	7 1 3 4 5	5 6 2 0 7	5 1 9 7 1	1 6 2 1 9
5 3 8 3 1	0 9 7 6 5	3 5 4 3 2	6 3 4 2 1	3 9 0 6 0
6 9 8 6 2	2 4 9 7 8	8 5 7 1 0	9 4 1 4 1	1 9 7 4 6
0 8 0 8 4	7 9 5 8 0	4 4 3 4 4	4 2 8 0 4	1 7 2 4 3
7 9 6 9 6	7 2 4 3 5	5 5 0 9 6	4 6 2 1 8	8 0 9 0 7
8 3 8 4 6	4 2 2 2 0	4 8 4 2 0	4 0 5 2 4	3 5 2 8 8
5 9 9 5 8	3 8 1 2 2	3 7 0 5 9	9 2 7 9 7	7 6 0 7 0
7 1 9 1 1	2 4 7 7 1	9 3 5 4 5	3 4 7 2 3	3 4 5 1 3
4 1 8 3 9	8 7 5 6 0	2 3 5 2 0	6 1 0 0 1	7 8 5 8 7
5 7 0 4 9	4 1 2 2 0	8 2 9 7 7	5 5 7 3 1	1 5 3 3 2
8 9 9 7 0	0 3 2 8 4	2 9 9 5 1	1 1 9 8 5	2 2 1 8 7
8 4 7 8 9	3 5 6 9 2	5 9 5 6 5	2 6 7 9 8	4 6 4 6 1
3 1 8 5 3	0 4 7 3 4	8 3 9 9 3	6 6 0 1 3	5 7 4 4 5
1 6 7 1 0	4 2 3 2 0	7 7 5 7 0	3 5 9 8 7	1 9 6 7 2
1 7 1 1 3	2 7 9 0 5	5 6 9 1 8	1 2 3 6 9	6 6 6 9 6
1 8 1 0 7	9 6 8 1 5	1 4 2 9 8	7 9 9 3 2	9 4 8 1 8
1 9 5 3 6	3 0 1 5 8	1 4 1 2 8	7 1 4 5 5	6 6 0 7 8
3 1 3 8 5	0 1 5 0 5	6 4 2 6 1	5 6 2 6 5	2 6 1 7 4
2 4 3 7 9	5 7 8 7 2	1 5 0 8 9	3 8 3 2 0	9 8 2 0 3
2 5 8 5 8	6 9 2 1 4	1 3 8 9 1	5 0 8 3 7	6 4 3 4 5
3 4 0 5 8	0 3 4 1 7	5 2 0 2 6	7 4 1 9 1	6 4 0 4 4
1 3 9 0 9	0 9 2 5 0	2 1 6 7 6	9 0 3 5 7	8 9 6 5 8
6 8 5 7 4	9 2 7 7 1	2 8 1 9 2	1 2 4 1 5	7 8 9 1 5
4 5 7 5 1	2 7 9 7 1	2 0 1 4 1	7 6 6 9 3	7 4 1 0 2
3 7 1 7 5	6 8 9 0 4	4 9 3 7 0	1 8 6 6 2	1 9 9 1 1
0 6 5 6 0	8 6 7 5 8	6 8 3 9 6	0 7 0 0 3	1 0 5 5 6

Table VIII
Table of Logarithms to Base 10

	0·00	0·02	0·04	0·06	0·08		0·00	0·02	0·04	0·06	0·08
1·0	0·000	0·009	0·017	0·025	0·033	5·5	0·740	0·742	0·744	0·745	0·747
1·1	0·041	0·049	0·057	0·064	0·072	5·6	0·748	0·750	0·751	0·753	0·754
1·2	0·079	0·086	0·093	0·100	0·107	5·7	0·756	0·757	0·759	0·760	0·762
1·3	0·114	0·121	0·127	0·134	0·140	5·8	0·763	0·765	0·766	0·768	0·769
1·4	0·146	0·152	0·158	0·164	0·170	5·9	0·771	0·772	0·774	0·775	0·777
1·5	0·176	0·182	0·188	0·193	0·199	6·0	0·778	0·780	0·781	0·782	0·784
1·6	0·204	0·210	0·215	0·220	0·225	6·1	0·785	0·787	0·788	0·790	0·791
1·7	0·230	0·236	0·241	0·246	0·250	6·2	0·792	0·794	0·795	0·797	0·798
1·8	0·255	0·260	0·265	0·270	0·274	6·3	0·799	0·801	0·802	0·803	0·805
1·9	0·279	0·283	0·288	0·292	0·297	6·4	0·806	0·808	0·809	0·810	0·812
2·0	0·301	0·305	0·310	0·314	0·318	6·5	0·813	0·814	0·816	0·817	0·818
2·1	0·322	0·326	0·330	0·334	0·338	6·6	0·820	0·821	0·822	0·823	0·825
2·2	0·342	0·346	0·350	0·354	0·358	6·7	0·826	0·827	0·829	0·830	0·831
2·3	0·362	0·365	0·369	0·373	0·377	6·8	0·833	0·834	0·835	0·836	0·838
2·4	0·380	0·384	0·387	0·391	0·394	6·9	0·839	0·840	0·841	0·843	0·844
2·5	0·398	0·401	0·405	0·408	0·412	7·0	0·845	0·846	0·848	0·849	0·850
2·6	0·415	0·418	0·422	0·425	0·428	7·1	0·851	0·852	0·854	0·855	0·856
2·7	0·431	0·435	0·438	0·441	0·444	7·2	0·857	0·859	0·860	0·861	0·862
2·8	0·447	0·450	0·453	0·456	0·459	7·3	0·863	0·865	0·866	0·867	0·868
2·9	0·462	0·465	0·468	0·471	0·474	7·4	0·869	0·870	0·872	0·873	0·874
3·0	0·477	0·480	0·483	0·486	0·489	7·5	0·875	0·876	0·877	0·879	0·880
3·1	0·491	0·494	0·497	0·500	0·502	7·6	0·881	0·882	0·883	0·884	0·885
3·2	0·505	0·508	0·511	0·513	0·516	7·7	0·886	0·888	0·889	0·890	0·891
3·3	0·519	0·521	0·524	0·526	0·529	7·8	0·892	0·893	0·894	0·895	0·897
3·4	0·531	0·534	0·537	0·539	0·542	7·9	0·898	0·899	0·900	0·901	0·902
3·5	0·544	0·547	0·549	0·551	0·554	8·0	0·903	0·904	0·905	0·906	0·907
3·6	0·556	0·559	0·561	0·563	0·566	8·1	0·908	0·910	0·911	0·912	0·913
3·7	0·568	0·571	0·573	0·575	0·577	8·2	0·914	0·915	0·916	0·917	0·918
3·8	0·580	0·582	0·584	0·587	0·589	8·3	0·919	0·920	0·921	0·922	0·923
3·9	0·591	0·593	0·595	0·598	0·600	8·4	0·924	0·925	0·926	0·927	0·928
4·0	0·602	0·604	0·606	0·609	0·611	8·5	0·929	0·930	0·931	0·932	0·933
4·1	0·613	0·615	0·617	0·619	0·621	8·6	0·934	0·936	0·937	0·938	0·939
4·2	0·623	0·625	0·627	0·629	0·631	8·7	0·940	0·941	0·942	0·943	0·943
4·3	0·633	0·635	0·637	0·639	0·641	8·8	0·944	0·945	0·946	0·947	0·948
4·4	0·643	0·645	0·647	0·649	0·651	8·9	0·949	0·950	0·951	0·952	0·953
4·5	0·653	0·655	0·657	0·659	0·661	9·0	0·954	0·955	0·956	0·957	0·958
4·6	0·663	0·665	0·667	0·668	0·670	9·1	0·959	0·960	0·961	0·962	0·963
4·7	0·672	0·674	0·676	0·678	0·679	9·2	0·964	0·965	0·966	0·967	0·968
4·8	0·681	0·683	0·685	0·687	0·688	9·3	0·968	0·969	0·970	0·971	0·972
4·9	0·690	0·692	0·694	0·695	0·697	9·4	0·973	0·974	0·975	0·976	0·977
5·0	0·699	0·701	0·702	0·704	0·706	9·5	0·978	0·979	0·980	0·980	0·981
5·1	0·708	0·709	0·711	0·713	0·714	9·6	0·982	0·983	0·984	0·985	0·986
5·2	0·716	0·718	0·719	0·721	0·723	9·7	0·987	0·988	0·989	0·989	0·990
5·3	0·724	0·726	0·728	0·729	0·731	9·8	0·991	0·992	0·993	0·994	0·995
5·4	0·732	0·734	0·736	0·737	0·739	9·9	0·996	0·997	0·997	0·998	0·999

Table VIII (continued)
Table of Natural Logarithms

	0·00	0·02	0·04	0·06	0·08		0·00	0·02	0·04	0·06	0·08
1·0	0·000	0·020	0·039	0·058	0·077	5·5	1·705	1·708	1·712	1·716	1·719
1·1	0·095	0·113	0·131	0·148	0·166	5·6	1·723	1·726	1·730	1·733	1·737
1·2	0·182	0·199	0·215	0·231	0·247	5·7	1·740	1·744	1·747	1·751	1·754
1·3	0·262	0·278	0·293	0·307	0·322	5·8	1·758	1·761	1·765	1·768	1·772
1·4	0·336	0·351	0·365	0·378	0·392	5·9	1·775	1·778	1·782	1·785	1·788
1·5	0·405	0·419	0·432	0·445	0·457	6·0	1·792	1·795	1·798	1·802	1·805
1·6	0·470	0·482	0·495	0·507	0·519	6·1	1·808	1·812	1·815	1·818	1·821
1·7	0·531	0·542	0·554	0·565	0·577	6·2	1·825	1·828	1·831	1·834	1·837
1·8	0·588	0·599	0·610	0·621	0·631	6·3	1·841	1·844	1·847	1·850	1·853
1·9	0·642	0·652	0·663	0·673	0·683	6·4	1·856	1·859	1·863	1·866	1·869
2·0	0·693	0·703	0·713	0·723	0·732	6·5	1·872	1·875	1·878	1·881	1·884
2·1	0·742	0·751	0·761	0·770	0·779	6·6	1·887	1·890	1·893	1·896	1·899
2·2	0·788	0·798	0·806	0·815	0·824	6·7	1·902	1·905	1·908	1·911	1·914
2·3	0·833	0·842	0·850	0·859	0·867	6·8	1·917	1·920	1·923	1·926	1·929
2·4	0·875	0·884	0·892	0·900	0·908	6·9	1·932	1·934	1·937	1·940	1·943
2·5	0·916	0·924	0·932	0·940	0·948	7·0	1·946	1·949	1·952	1·954	1·957
2·6	0·956	0·963	0·971	0·978	0·986	7·1	1·960	1·963	1·966	1·969	1·971
2·7	0·993	1·001	1·008	1·015	1·022	7·2	1·974	1·977	1·980	1·982	1·985
2·8	1·030	1·037	1·044	1·051	1·058	7·3	1·988	1·991	1·993	1·996	1·999
2·9	1·065	1·072	1·078	1·085	1·092	7·4	2·001	2·004	2·007	2·010	2·012
3·0	1·099	1·105	1·112	1·118	1·125	7·5	2·015	2·018	2·020	2·023	2·026
3·1	1·131	1·138	1·144	1·151	1·157	7·6	2·028	2·031	2·033	2·036	2·039
3·2	1·163	1·169	1·176	1·182	1·188	7·7	2·041	2·044	2·046	2·049	2·052
3·3	1·194	1·200	1·206	1·212	1·218	7·8	2·054	2·057	2·059	2·062	2·064
3·4	1·224	1·230	1·235	1·241	1·247	7·9	2·067	2·069	2·072	2·074	2·077
3·5	1·253	1·258	1·264	1·270	1·275	8·0	2·079	2·082	2·084	2·087	2·089
3·6	1·281	1·286	1·292	1·297	1·303	8·1	2·092	2·094	2·097	2·099	2·102
3·7	1·308	1·314	1·319	1·324	1·330	8·2	2·104	2·107	2·109	2·111	2·114
3·8	1·335	1·340	1·345	1·351	1·356	8·3	2·116	2·119	2·121	2·123	2·126
3·9	1·361	1·366	1·371	1·376	1·381	8·4	2·128	2·131	2·133	2·135	2·138
4·0	1·386	1·391	1·396	1·401	1·406	8·5	2·140	2·142	2·145	2·147	2·149
4·1	1·411	1·416	1·421	1·426	1·430	8·6	2·152	2·154	2·156	2·159	2·161
4·2	1·435	1·440	1·445	1·449	1·454	8·7	2·163	2·166	2·168	2·170	2·172
4·3	1·459	1·463	1·468	1·472	1·477	8·8	2·175	2·177	2·179	2·182	2·184
4·4	1·482	1·486	1·491	1·495	1·500	8·9	2·186	2·188	2·191	2·193	2·195
4·5	1·504	1·509	1·513	1·517	1·522	9·0	2·197	2·199	2·202	2·204	2·206
4·6	1·526	1·530	1·535	1·539	1·543	9·1	2·208	2·210	2·213	2·215	2·217
4·7	1·548	1·552	1·556	1·560	1·564	9·2	2·219	2·221	2·224	2·226	2·228
4·8	1·569	1·573	1·577	1·581	1·585	9·3	2·230	2·232	2·234	2·236	2·239
4·9	1·589	1·593	1·597	1·601	1·605	9·4	2·241	2·243	2·245	2·247	2·249
5·0	1·609	1·613	1·617	1·621	1·625	9·5	2·251	2·253	2·255	2·258	2·260
5·1	1·629	1·633	1·637	1·641	1·645	9·6	2·262	2·264	2·266	2·268	2·270
5·2	1·649	1·652	1·656	1·660	1·664	9·7	2·272	2·274	2·276	2·278	2·280
5·3	1·668	1·671	1·675	1·679	1·683	9·8	2·282	2·284	2·286	2·288	2·291
5·4	1·686	1·690	1·694	1·697	1·701	9·9	2·293	2·295	2·297	2·299	2·301

For each tenfold increase beyond the range of this table, add 2·303.
For each hundredfold increase beyond the range of this table, add 4·605.

Table IX

Values of $\sin^{-1}\sqrt{p}$

p	0·00	0·01	0·02	0·03	0·04	0·05	0·06	0·07	0·08	0·09
0·0	0·00	0·10	0·14	0·17	0·20	0·23	0·25	0·27	0·29	0·30
0·1	0·32	0·34	0·35	0·37	0·38	0·40	0·41	0·42	0·44	0·45
0·2	0·46	0·48	0·49	0·50	0·51	0·52	0·54	0·55	0·56	0·57
0·3	0·58	0·59	0·60	0·61	0·62	0·63	0·64	0·65	0·66	0·67
0·4	0·68	0·69	0·71	0·72	0·73	0·74	0·75	0·76	0·77	0·78
0·5	0·79	0·80	0·81	0·82	0·83	0·84	0·85	0·86	0·87	0·88
0·6	0·89	0·90	0·91	0·92	0·93	0·94	0·95	0·96	0·97	0·98
0·7	0·99	1·00	1·01	1·02	1·04	1·05	1·06	1·07	1·08	1·09
0·8	1·11	1·12	1·13	1·15	1·16	1·17	1·19	1·20	1·22	1·23
0·9	1·25	1·27	1·28	1·30	1·32	1·35	1·37	1·40	1·43	1·47

Table X

Values of $\sinh^{-1}\sqrt{x}$

x	0·0	0·1	0·2	0·3	0·4	0·5	0·6	0·7	0·8	0·9
$\sinh^{-1}\sqrt{x}$	0·00	0·31	0·43	0·52	0·60	0·66	0·71	0·76	0·80	0·84

x	0	1	2	3	4	5	6	7	8	9
0	0·00	0·88	1·15	1·32	1·44	1·54	1·63	1·70	1·76	1·82
10	1·87	1·91	1·96	1·99	2·03	2·06	2·09	2·12	2·15	2·18
20	2·00	2·23	2·25	2·27	2·29	2·31	2·33	2·35	2·37	2·39
30	2·40	2·42	2·43	2·45	2·46	2·48	2·49	2·51	2·52	2·53
40	2·54	2·56	2·57	2·58	2·59	2·60	2·61	2·62	2·63	2·64
50	2·65	2·66	2·67	2·68	2·69	2·70	2·71	2·72	2·73	2·74
60	2·74	2·75	2·76	2·77	2·78	2·78	2·79	2·80	2·81	2·81
70	2·82	2·83	2·83	2·84	2·85	2·86	2·86	2·87	2·87	2·88
80	2·89	2·89	2·90	2·91	2·91	2·92	2·92	2·93	2·93	2·94
90	2·95	2·95	2·96	2·96	2·97	2·97	2·98	2·98	2·99	2·99

For each tenfold increase in x beyond the range of this table add 1·15 to the value of $\sinh^{-1}\sqrt{x}$. For example, $\sinh^{-1}\sqrt{210} = 2{\cdot}23 + 1{\cdot}15 = 3{\cdot}38$ and $\sinh^{-1}\sqrt{2{\cdot}15} = 2{\cdot}24 + 1{\cdot}15 = 3{\cdot}39$.

Table XI
Table of Squares and Square Roots

n	n^2	$\sqrt{n}$	$\sqrt{(10n)}$	n	n^2	$\sqrt{n}$	$\sqrt{(10n)}$
1	1	1·000	3·162	51	2601	7·141	22·583
2	4	1·414	4·472	52	2704	7·211	22·804
3	9	1·732	5·477	53	2809	7·280	23·022
4	16	2·000	6·325	54	2916	7·348	23·238
5	25	2·236	7·071	55	3025	7·416	23·452
6	36	2·449	7·746	56	3136	7·483	23·664
7	49	2·646	8·367	57	3249	7·550	23·875
8	64	2·828	8·944	58	3364	7·616	24·083
9	81	3·000	9·487	59	3481	7·681	24·290
10	100	3·162	10·000	60	3600	7·746	24·495
11	121	3·317	10·488	61	3721	7·810	24·698
12	144	3·464	10·954	62	3844	7·874	24·900
13	169	3·606	11·402	63	3969	7·937	25·100
14	196	3·742	11·832	64	4096	8·000	25·298
15	225	3·873	12·247	65	4225	8·062	25·495
16	256	4·000	12·649	66	4356	8·124	25·690
17	289	4·123	13·038	67	4489	8·185	25·884
18	324	4·243	13·416	68	4624	8·246	26·077
19	361	4·359	13·784	69	4761	8·307	26·268
20	400	4·472	14·142	70	4900	8·367	26·458
21	441	4·583	14·491	71	5041	8·426	26·646
22	484	4·690	14·832	72	5184	8·485	26·833
23	529	4·796	15·166	73	5329	8·544	27·019
24	576	4·899	15·492	74	5476	8·602	27·203
25	625	5·000	15·811	75	5625	8·660	27·386
26	676	5·099	16·125	76	5776	8·718	27·568
27	729	5·196	16·432	77	5929	8·775	27·749
28	784	5·292	16·733	78	6084	8·832	27·928
29	841	5·385	17·029	79	6241	8·888	28·107
30	900	5·477	17·321	80	6400	8·944	28·284
31	961	5·568	17·607	81	6561	9·000	28·460
32	1024	5·657	17·889	82	6724	9·055	28·636
33	1089	5·745	18·166	83	6889	9·110	28·810
34	1156	5·831	18·439	84	7056	9·165	28·983
35	1225	5·916	18·708	85	7225	9·220	29·155
36	1296	6·000	18·974	86	7396	9·274	29·326
37	1369	6·083	19·235	87	7569	9·327	29·496
38	1444	6·164	19·494	88	7744	9·381	29·665
39	1521	6·245	19·748	89	7921	9·434	29·833
40	1600	6·325	20·000	90	8100	9·487	30·000
41	1681	6·403	20·248	91	8281	9·539	30·166
42	1764	6·481	20·494	92	8464	9·592	30·332
43	1849	6·557	20·736	93	8649	9·644	30·496
44	1936	6·633	20·976	94	8836	9·695	30·659
45	2025	6·708	21·213	95	9025	9·747	30·822
46	2116	6·782	21·448	96	9216	9·798	30·984
47	2209	6·856	21·679	97	9409	9·849	31·145
48	2304	6·928	21·909	98	9604	9·899	31·305
49	2401	7·000	22·136	99	9801	9·950	31·464
50	2500	7·071	22·361	100	10000	10·000	31·623

Table XI (continued)

Table of Squares and Square Roots

n	n^2	$\sqrt{n}$	$\sqrt{(10n)}$	n	n^2	$\sqrt{n}$	$\sqrt{(10n)}$
101	10201	10·050	31·780	151	22801	12·288	38·859
102	10404	10·100	31·937	152	23104	12·329	38·987
103	10609	10·149	32·094	153	23409	12·369	39·115
104	10816	10·198	32·249	154	23716	12·410	39·243
105	11025	10·247	32·404	155	24025	12·450	39·370
106	11236	10·296	32·558	156	24336	12·490	39·497
107	11449	10·344	32·711	157	24649	12·530	39·623
108	11664	10·392	32·863	158	24964	12·570	39·749
109	11881	10·440	33·015	159	25281	12·610	39·875
110	12100	10·488	33·166	160	25600	12·649	40·000
111	12321	10·536	33·317	161	25921	12·689	40·125
112	12544	10·583	33·466	162	26244	12·728	40·249
113	12769	10·630	33·615	163	26569	12·767	40·373
114	12996	10·677	33·764	164	26896	12·806	40·497
115	13225	10·724	33·912	165	27225	12·845	40·620
116	13456	10·770	34·059	166	27556	12·884	40·743
117	13689	10·817	34·205	167	27889	12·923	40·866
118	13924	10·863	34·351	168	28224	12·961	40·988
119	14161	10·909	34·496	169	28561	13·000	41·110
120	14400	10·954	34·641	170	28900	13·038	41·231
121	14641	11·000	34·785	171	29241	13·077	41·352
122	14884	11·045	34·928	172	29584	13·115	41·473
123	15129	11·091	35·071	173	29929	13·153	41·593
124	15376	11·136	35·214	174	30276	13·191	41·713
125	15625	11·180	35·355	175	30625	13·229	41·833
126	15876	11·225	35·496	176	30976	13·266	41·952
127	16129	11·269	35·637	177	31329	13·304	42·071
128	16384	11·314	35·777	178	31684	13·342	42·190
129	16641	11·358	35·917	179	32041	13·379	42·308
130	16900	11·402	36·056	180	32400	13·416	42·426
131	17161	11·446	36·194	181	32761	13·454	42·544
132	17424	11·489	36·332	182	33124	13·491	42·661
133	17689	11·533	36·469	183	33489	13·528	42·778
134	17956	11·576	36·606	184	33856	13·565	42·895
135	18225	11·619	36·742	185	34225	13·601	43·012
136	18496	11·662	36·878	186	34596	13·638	43·128
137	18769	11·705	37·014	187	34969	13·675	43·243
138	19044	11·747	37·148	188	35344	13·711	43·359
139	19321	11·790	37·283	189	35721	13·748	43·474
140	19600	11·832	37·417	190	36100	13·784	43·589
141	19881	11·874	37·550	191	36481	13·820	43·704
142	20164	11·916	37·683	192	36864	13·856	43·818
143	20449	11·958	37·815	193	37249	13·892	43·932
144	20736	12·000	37·947	194	37636	13·928	44·045
145	21025	12·042	38·079	195	38025	13·964	44·159
146	21316	12·083	38·210	196	38416	14·000	44·272
147	21609	12·124	38·341	197	38809	14·036	44·385
148	21904	12·166	38·471	198	39204	14·071	44·497
149	22201	12·207	38·601	199	39601	14·107	44·609
150	22500	12·247	38·730	200	40000	14·142	44·721

Table XI (*continued*)

Table of Squares and Square Roots

n	n^2	$\sqrt{n}$	$\sqrt{(10n)}$	n	n^2	$\sqrt{n}$	$\sqrt{(10n)}$
201	40401	14·177	44·833	251	63001	15·843	50·100
202	40804	14·213	44·944	252	63504	15·875	50·200
203	41209	14·248	45·056	253	64009	15·906	50·299
204	41616	14·283	45·166	254	64516	15·937	50·398
205	42025	14·318	45·277	255	65025	15·969	50·498
206	42436	14·353	45·387	256	65536	16·000	50·596
207	42849	14·387	45·497	257	66049	16·031	50·695
208	43264	14·422	45·607	258	66564	16·062	50·794
209	43681	14·457	45·717	259	67081	16·093	50·892
210	44100	14·491	45·826	260	67600	16·125	50·990
211	44521	14·526	45·935	261	68121	16·155	51·088
212	44944	14·560	46·043	262	68644	16·186	51·186
213	45369	14·595	46·152	263	69169	16·217	51·284
214	45796	14·629	46·260	264	69696	16·248	51·381
215	46225	14·663	46·368	265	70225	16·279	51·478
216	46656	14·697	46·476	266	70756	16·310	51·575
217	47089	14·731	46·583	267	71289	16·340	51·672
218	47524	14·765	46·690	268	71824	16·371	51·769
219	47961	14·799	46·797	269	72361	16·401	51·865
220	48400	14·832	46·904	270	72900	16·432	51·962
221	48841	14·866	47·011	271	73441	16·462	52·058
222	49284	14·900	47·117	272	73984	16·492	52·154
223	49729	14·933	47·223	273	74529	16·523	52·249
224	50176	14·967	47·329	274	75076	16·553	52·345
225	50625	15·000	47·434	275	75625	16·583	52·440
226	51076	15·033	47·539	276	76176	16·613	52·536
227	51529	15·067	47·645	277	76729	16·643	52·631
228	51984	15·100	47·749	278	77284	16·673	52·726
229	52441	15·133	47·854	279	77841	16·703	52·820
230	52900	15·166	47·958	280	78400	16·733	52·915
231	53361	15·199	48·062	281	78961	16·763	53·009
232	53824	15·232	48·166	282	79524	16·793	53·104
233	54289	15·264	48·270	283	80089	16·823	53·198
234	54756	15·297	48·374	284	80656	16·852	53·292
235	55225	15·330	48·477	285	81225	16·882	53·385
236	55696	15·362	48·580	286	81796	16·912	53·479
237	56169	15·395	48·683	287	82369	16·941	53·572
238	56644	15·427	48·785	288	82944	16·971	53·666
239	57121	15·460	48·888	289	83521	17·000	53·759
240	57600	15·492	48·990	290	84100	17·029	53·852
241	58081	15·524	49·092	291	84681	17·059	53·944
242	58564	15·556	49·193	292	85264	17·088	54·037
243	59049	15·588	49·295	293	85849	17·117	54·129
244	59536	15·620	49·396	294	86436	17·146	54·222
245	60025	15·652	49·497	295	87025	17·176	54·314
246	60516	15·684	49·598	296	87616	17·205	54·406
247	61009	15·716	49·699	297	88209	17·234	54·498
248	61504	15·748	49·800	298	88804	17·263	54·589
249	62001	15·780	49·900	299	89401	17·292	54·681
250	62500	15·811	50·000	300	90000	17·321	54·772

Table XI (continued)

Table of Squares and Square Roots

n	n^2	$\sqrt{n}$	$\sqrt{(10n)}$	n	n^2	$\sqrt{n}$	$\sqrt{(10n)}$
301	90601	17·349	54·863	351	123201	18·735	59·245
302	91204	17·378	54·955	352	123904	18·762	59·330
303	91809	17·407	55·045	353	124609	18·788	59·414
304	92416	17·436	55·136	354	125316	18·815	59·498
305	93025	17·464	55·227	355	126025	18·841	59·582
306	93636	17·493	55·317	356	126736	18·868	59·666
307	94249	17·521	55·408	357	127449	18·894	59·749
308	94864	17·550	55·498	358	128164	18·921	59·833
309	95481	17·578	55·588	359	128881	18·947	59·917
310	96100	17·607	55·678	360	129600	18·974	60·000
311	96721	17·635	55·767	361	130321	19·000	60·083
312	97344	17·664	55·857	362	131044	19·026	60·166
313	97969	17·692	55·946	363	131769	19·053	60·249
314	98596	17·720	56·036	364	132496	19·079	60·332
315	99225	17·748	56·125	365	133225	19·105	60·415
316	99856	17·776	56·214	366	133956	19·131	60·498
317	100489	17·804	56·303	367	134689	19·157	60·581
318	101124	17·833	56·391	368	135424	19·183	60·663
319	101761	17·861	56·480	369	136161	19·209	60·745
320	102400	17·889	56·569	370	136900	19·235	60·828
321	103041	17·916	56·657	371	137641	19·261	60·910
322	103684	17·944	56·745	372	138384	19·287	60·992
323	104329	17·972	56·833	373	139129	19·313	61·074
324	104976	18·000	56·921	374	139876	19·339	61·156
325	105625	18·028	57·009	375	140625	19·365	61·237
326	106276	18·055	57·096	376	141376	19·391	61·319
327	106929	18·083	57·184	377	142129	19·416	61·400
328	107584	18·111	57·271	378	142884	19·442	61·482
329	108241	18·138	57·359	379	143641	19·468	61·563
330	108900	18·166	57·446	380	144400	19·494	61·644
331	109561	18·193	57·533	381	145161	19·519	61·725
332	110224	18·221	57·619	382	145924	19·545	61·806
333	110889	18·248	57·706	383	146689	19·570	61·887
334	111556	18·276	57·793	384	147456	19·596	61·968
335	112225	18·303	57·879	385	148225	19·621	62·048
336	112896	18·330	57·966	386	148996	19·647	62·129
337	113569	18·358	58·052	387	149769	19·672	62·209
338	114244	18·385	58·138	388	150544	19·698	62·290
339	114921	18·412	58·224	389	151321	19·723	62·370
340	115600	18·439	58·310	390	152100	19·748	62·450
341	116281	18·466	58·395	391	152881	19·774	62·530
342	116964	18·493	58·481	392	153664	19·799	62·610
343	117649	18·520	58·566	393	154449	19·824	62·690
344	118336	18·547	58·652	394	155236	19·849	62·769
345	119025	18·574	58·737	395	156025	19·875	62·849
346	119716	18·601	58·822	396	156816	19·900	62·929
347	120409	18·628	58·907	397	157609	19·925	63·008
348	121104	18·655	58·992	398	158404	19·950	63·087
349	121801	18·682	59·076	399	159201	19·975	63·166
350	122500	18·708	59·161	400	160000	20·000	63·246

Table XI (continued)

Table of Squares and Square Roots

n	n^2	$\sqrt{n}$	$\sqrt{(10n)}$	n	n^2	$\sqrt{n}$	$\sqrt{(10n)}$
401	160801	20·025	63·325	451	203401	21·237	67·157
402	161604	20·050	63·403	452	204304	21·260	67·231
403	162409	20·075	63·482	453	205209	21·284	67·305
404	163216	20·100	63·561	454	206116	21·307	67·380
405	164025	20·125	63·640	455	207025	21·331	67·454
406	164836	20·149	63·718	456	207936	21·354	67·528
407	165649	20·174	63·797	457	208849	21·378	67·602
408	166464	20·199	63·875	458	209764	21·401	67·676
409	167281	20·224	63·953	459	210681	21·424	67·750
410	168100	20·248	64·031	460	211600	21·448	67·823
411	168921	20·273	64·109	461	212521	21·471	67·897
412	169744	20·298	64·187	462	213444	21·494	67·971
413	170569	20·322	64·265	463	214369	21·517	68·044
414	171396	20·347	64·343	464	215296	21·541	68·118
415	172225	20·372	64·420	465	216225	21·564	68·191
416	173056	20·396	64·498	466	217156	21·587	68·264
417	173889	20·421	64·576	467	218089	21·610	68·337
418	174724	20·445	64·653	468	219024	21·633	68·411
419	175561	20·469	64·730	469	219961	21·656	68·484
420	176400	20·494	64·807	470	220900	21·679	68·557
421	177241	20·518	64·885	471	221841	21·703	68·629
422	178084	20·543	64·962	472	222784	21·726	68·702
423	178929	20·567	65·038	473	223729	21·749	68·775
424	179776	20·591	65·115	474	224676	21·772	68·848
425	180625	20·616	65·192	475	225625	21·794	68·920
426	181476	20·640	65·269	476	226576	21·817	68·993
427	182329	20·664	65·345	477	227529	21·840	69·065
428	183184	20·688	65·422	478	228484	21·863	69·138
429	184041	20·712	65·498	479	229441	21·886	69·210
430	184900	20·736	65·574	480	230400	21·909	69·282
431	185761	20·761	65·651	481	231361	21·932	69·354
432	186624	20·785	65·727	482	232324	21·954	69·426
433	187489	20·809	65·803	483	233289	21·977	69·498
434	188356	20·833	65·879	484	234256	22·000	69·570
435	189225	20·857	65·955	485	235225	22·023	69·642
436	190096	20·881	66·030	486	236196	22·045	69·714
437	190969	20·905	66·106	487	237169	22·068	69·785
438	191844	20·928	66·182	488	238144	22·091	69·857
439	192721	20·952	66·257	489	239121	22·113	69·929
440	193600	20·976	66·332	490	240100	22·136	70·000
441	194481	21·000	66·408	491	241081	22·159	70·071
442	195364	21·024	66·483	492	242064	22·181	70·143
443	196249	21·048	66·558	493	243049	22·204	70·214
444	197136	21·071	66·633	494	244036	22·226	70·285
445	198025	21·095	66·708	495	245025	22·249	70·356
446	198916	21·119	66·783	496	246016	22·271	70·427
447	199809	21·142	66·858	497	247009	22·293	70·498
448	200704	21·166	66·933	498	248004	22·316	70·569
449	201601	21·190	67·007	499	249001	22·338	70·640
450	202500	21·213	67·082	500	250000	22·361	70·711

Table XI (continued)

Table of Squares and Square Roots

n	n^2	$\sqrt{n}$	$\sqrt{(10n)}$	n	n^2	$\sqrt{n}$	$\sqrt{(10n)}$
501	251001	22·383	70·781	551	303601	23·473	74·229
502	252004	22·405	70·852	552	304704	23·495	74·297
503	253009	22·428	70·922	553	305809	23·516	74·364
504	254016	22·450	70·993	554	306916	23·537	74·431
505	255025	22·472	71·063	555	308025	23·558	74·498
506	256036	22·494	71·134	556	309136	23·580	74·565
507	257049	22·517	71·204	557	310249	23·601	74·632
508	258064	22·539	71·274	558	311364	23·622	74·699
509	259081	22·561	71·344	559	312481	23·643	74·766
510	260100	22·583	71·414	560	313600	23·664	74·833
511	261121	22·605	71·484	561	314721	23·685	74·900
512	262144	22·627	71·554	562	315844	23·707	74·967
513	263169	22·650	71·624	563	316969	23·728	75·033
514	264196	22·672	71·694	564	318096	23·749	75·100
515	265225	22·694	71·764	565	319225	23·770	75·166
516	266256	22·716	71·833	566	320356	23·791	75·233
517	267289	22·738	71·903	567	321489	23·812	75·299
518	268324	22·760	71·972	568	322624	23·833	75·366
519	269361	22·782	72·042	569	323761	23·854	75·432
520	270400	22·804	72·111	570	324900	23·875	75·498
521	271441	22·825	72·180	571	326041	23·896	75·565
522	272484	22·847	72·250	572	327184	23·917	75·631
523	273529	22·869	72·319	573	328329	23·937	75·697
524	274576	22·891	72·388	574	329476	23·958	75·763
525	275625	22·913	72·457	575	330625	23·979	75·829
526	276676	22·935	72·526	576	331776	24·000	75·895
527	277729	22·956	72·595	577	332929	24·021	75·961
528	278784	22·978	72·664	578	334084	24·042	76·026
529	279841	23·000	72·732	579	335241	24·062	76·092
530	280900	23·022	72·801	580	336400	24·083	76·158
531	281961	23·043	72·870	581	337561	24·104	76·223
532	283024	23·065	72·938	582	338724	24·125	76·289
533	284089	23·087	73·007	583	339889	24·145	76·354
534	285156	23·108	73·075	584	341056	24·166	76·420
535	286225	23·130	73·144	585	342225	24·187	76·485
536	287296	23·152	73·212	586	343396	24·207	76·551
537	288369	23·173	73·280	587	344569	24·228	76·616
538	289444	23·195	73·348	588	345744	24·249	76·681
539	290521	23·216	73·417	589	346921	24·269	76·746
540	291600	23·238	73·485	590	348100	24·290	76·811
541	292681	23·259	73·553	591	349281	24·310	76·877
542	293764	23·281	73·621	592	350464	24·331	76·942
543	294849	23·302	73·689	593	351649	24·352	77·006
544	295936	23·324	73·756	594	352836	24·372	77·071
545	297025	23·345	73·824	595	354025	24·393	77·136
546	298116	23·367	73·892	596	355216	24·413	77·201
547	299209	23·388	73·959	597	356409	24·434	77·266
548	300304	23·409	74·027	598	357604	24·454	77·330
549	301401	23·431	74·095	599	358801	24·474	77·395
550	302500	23·452	74·162	600	360000	24·495	77·460

Table XI (continued)

Table of Squares and Square Roots

n	n^2	$\sqrt{n}$	$\sqrt{(10n)}$	n	n^2	$\sqrt{n}$	$\sqrt{(10n)}$
601	361201	24·515	77·524	651	423801	25·515	80·685
602	362404	24·536	77·589	652	425104	25·534	80·747
603	363609	24·556	77·653	653	426409	25·554	80·808
604	364816	24·576	77·717	654	427716	25·573	80·870
605	366025	24·597	77·782	655	429025	25·593	80·932
606	367236	24·627	77·846	656	430336	25·612	80·994
607	368449	24·637	77·910	657	431649	25·632	81·056
608	369664	24·658	77·974	658	432964	25·652	81·117
609	370881	24·678	78·038	659	434281	25·671	81·179
610	372100	24·698	78·102	660	435600	25·690	81·240
611	373321	24·718	78·166	661	436921	25·710	81·302
612	374544	24·739	78·230	662	438244	25·729	81·363
613	375769	24·759	78·294	663	439569	25·749	81·425
614	376996	24·779	78·358	664	440896	25·768	81·486
615	378225	24·799	78·422	665	442225	25·788	81·548
616	379456	24·819	78·486	666	443556	25·807	81·609
617	380689	24·839	78·549	667	444889	25·826	81·670
618	381924	24·860	78·613	668	446224	25·846	81·731
619	383161	24·880	78·677	669	447561	25·865	81·792
620	384400	24·900	78·740	670	448900	25·884	81·854
621	385641	24·920	78·804	671	450241	25·904	81·915
622	386884	24·940	78·867	672	451584	25·923	81·976
623	388129	24·960	78·930	673	452929	25·942	82·037
624	389376	24·980	78·994	674	454276	25·962	82·098
625	390625	25·000	79·057	675	455625	25·981	82·158
626	391876	25·020	79·120	676	456976	26·000	82·219
627	393129	25·040	79·183	677	458329	26·019	82·280
628	394384	25·060	79·246	678	459684	26·038	82·341
629	395641	25·080	79·310	679	461041	26·058	82·401
630	396900	25·100	79·373	680	462400	26·077	82·462
631	398161	25·120	79·436	681	463761	26·096	82·523
632	399424	25·140	79·498	682	465124	26·115	82·583
633	400689	25·159	79·561	683	466489	26·134	82·644
634	401956	25·179	79·624	684	467856	26·153	82·704
635	403225	25·199	79·687	685	469225	26·173	82·765
636	404496	25·219	79·750	686	470596	26·192	82·825
637	405769	25·239	79·812	687	471969	26·211	82·885
638	407044	25·259	79·875	688	473344	26·230	82·946
639	408321	25·278	79·937	689	474721	26·249	83·006
640	409600	25·298	80·000	690	476100	26·268	83·066
641	410881	25·318	80·062	691	477481	26·287	83·126
642	412164	25·338	80·125	692	478864	26·306	83·187
643	413449	25·357	80·187	693	480249	26·325	83·247
644	414736	25·377	80·250	694	481636	26·344	83·307
645	416025	25·397	80·312	695	483025	26·363	83·367
646	417316	25·417	80·374	696	484416	26·382	83·427
647	418609	25·436	80·436	697	485809	26·401	83·487
648	419904	25·456	80·498	698	487204	26·420	83·546
649	421201	25·475	80·561	699	488601	26·439	83·606
650	422500	25·495	80·623	700	490000	26·458	83·666

Table XI (continued)

Table of Squares and Square Roots

n	n^2	$\sqrt{n}$	$\sqrt{(10n)}$	n	n^2	$\sqrt{n}$	$\sqrt{(10n)}$
701	491401	26·476	83·726	751	564001	27·404	86·660
702	492804	26·495	83·785	752	565504	27·423	86·718
703	494209	26·514	83·845	753	567009	27·441	86·776
704	495616	26·533	83·905	754	568516	27·459	86·833
705	497025	26·552	83·964	755	570025	27·477	86·891
706	498436	26·571	84·024	756	571536	27·495	86·948
707	499849	26·589	84·083	757	573049	27·514	87·006
708	501264	26·608	84·143	758	574564	27·532	87·063
709	502681	26·627	84·202	759	576081	27·550	87·121
710	504100	26·646	84·261	760	577600	27·568	87·178
711	505521	26·665	84·321	761	579121	27·586	87·235
712	506944	26·683	84·380	762	580644	27·604	87·293
713	508369	26·702	84·439	763	582169	27·622	87·350
714	509796	26·721	84·499	764	583696	27·641	87·407
715	511225	26·739	84·558	765	585225	27·659	87·464
716	512656	26·758	84·617	766	586756	27·677	87·521
717	514089	26·777	84·676	767	588289	27·695	87·579
718	515524	26·796	84·735	768	589824	27·713	87·636
719	516961	26·814	84·794	769	591361	27·731	87·693
720	518400	26·833	84·853	770	592900	27·749	87·750
721	519841	26·851	84·912	771	594441	27·767	87·807
722	521284	26·870	84·971	772	595984	27·785	87·864
723	522729	26·889	85·029	773	597529	27·803	87·920
724	524176	26·907	85·088	774	599076	27·821	87·977
725	525625	26·926	85·147	775	600625	27·839	88·034
726	527076	26·944	85·206	776	602176	27·857	88·091
727	528529	26·963	85·264	777	603729	27·875	88·148
728	529984	26·981	85·323	778	605284	27·893	88·204
729	531441	27·000	85·381	779	606841	27·911	88·261
730	532900	27·019	85·440	780	608400	27·928	88·318
731	534361	27·037	85·499	781	609961	27·946	88·374
732	535824	27·055	85·557	782	611524	27·964	88·431
733	537289	27·074	85·615	783	613089	27·982	88·487
734	538756	27·092	85·674	784	614656	28·000	88·544
735	540225	27·111	85·732	785	616225	28·018	88·600
736	541696	27·129	85·790	786	617796	28·036	88·657
737	543169	27·148	85·849	787	619369	28·054	88·713
738	544644	27·166	85·907	788	620944	28·071	88·769
739	546121	27·185	85·965	789	622521	28·089	88·826
740	547600	27·203	86·023	790	624100	28·107	88·882
741	549081	27·221	86·081	791	625681	28·125	88·938
742	550564	27·240	86·139	792	627264	28·142	88·994
743	552049	27·258	86·197	793	628849	28·160	89·051
744	553536	27·276	86·255	794	630436	28·178	89·107
745	555025	27·295	86·313	795	632025	28·196	89·163
746	556516	27·313	86·371	796	633616	28·213	89·219
747	558009	27·331	86·429	797	635209	28·231	89·275
748	559504	27·350	86·487	798	636804	28·249	89·331
749	561001	27·368	86·545	799	638401	28·267	89·387
750	562500	27·386	86·603	800	640000	28·284	89·443

Table XI (*continued*)

Table of Squares and Square Roots

n	n^2	$\surd n$	$\surd(10n)$	n	n^2	$\surd n$	$\surd(10n)$
801	641601	28·302	89·499	851	724201	29·172	92·250
802	643204	28·320	89·554	852	725904	29·189	92·304
803	644809	28·337	89·610	853	727609	29·206	92·358
804	646416	28·355	89·666	854	729316	29·223	92·412
805	648025	28·373	89·722	855	731025	29·240	92·466
806	649636	28·390	89·778	856	732736	29·257	92·520
807	651249	28·408	89·833	857	734449	29·275	92·574
808	652864	28·425	89·889	858	736164	29·292	92·628
809	654481	28·443	89·944	859	737881	29·309	92·682
810	656100	28·460	90·000	860	739600	29·326	92·736
811	657721	28·478	90·056	861	741321	29·343	92·790
812	659344	28·496	90·111	862	743044	29·360	92·844
813	660969	28·513	90·167	863	744769	29·377	92·898
814	662596	28·531	90·222	864	746496	29·394	92·952
815	664225	28·548	90·277	865	748225	29·411	93·005
816	665856	28·566	90·333	866	749956	29·428	93·059
817	667489	28·583	90·388	867	751689	29·445	93·113
818	669124	28·601	90·443	868	753424	29·462	93·167
819	670761	28·618	90·499	869	755161	29·479	93·220
820	672400	28·636	90·554	870	756900	29·496	93·274
821	674041	28·653	90·609	871	758641	29·513	93·327
822	675684	28·671	90·664	872	760384	29·530	93·381
823	677329	28·688	90·719	873	762129	29·547	93·434
824	678976	28·705	90·774	874	763876	29·563	93·488
825	680625	28·723	90·830	875	765625	29·580	93·541
826	682276	28·740	90·885	876	767376	29·597	93·595
827	683929	28·758	90·940	877	769129	29·614	93·648
828	685584	28·775	90·995	878	770884	29·631	93·702
829	687241	28·792	91·049	879	772641	29·648	93·755
830	688900	28·810	91·104	880	774400	29·665	93·808
831	690561	28·827	91·159	881	776161	29·682	93·862
832	692224	28·844	91·214	882	777924	29·698	93·915
833	693889	28·862	91·269	883	779689	29·715	93·968
834	695556	28·879	91·324	884	781456	29·732	94·021
835	697225	28·896	91·378	885	783225	29·749	94·074
836	698896	28·914	91·433	886	784996	29·766	94·128
837	700569	28·931	91·488	887	786769	29·783	94·181
838	702244	28·948	91·542	888	788544	29·799	94·234
839	703921	28·965	91·597	889	790321	29·816	94·287
840	705600	28·983	91·652	890	792100	29·833	94·340
841	707281	29·000	91·706	891	793881	29·850	94·393
842	708964	29·017	91·761	892	795664	29·866	94·446
843	710649	29·034	91·815	893	797449	29·883	94·499
844	712336	29·052	91·869	894	799236	29·900	94·552
845	714025	29·069	91·924	895	801025	29·917	94·604
846	715716	29·086	91·978	896	802816	29·933	94·657
847	717409	29·103	92·033	897	804609	29·950	94·710
848	719104	29·120	92·087	898	806404	29·967	94·763
849	720801	29·138	92·141	899	808201	29·983	94·816
850	722500	29·155	92·195	900	810000	30·000	94·868

Table XI (continued)
Table of Squares and Square Roots

n	n^2	$\sqrt{n}$	$\sqrt{(10n)}$	n	n^2	$\sqrt{n}$	$\sqrt{(10n)}$
901	811801	30·017	94·921	951	904401	30·838	97·519
902	813604	30·033	94·974	952	906304	30·854	97·570
903	815409	30·050	95·026	953	908209	30·871	97·622
904	817216	30·067	95·079	954	910116	30·887	97·673
905	819025	30·083	95·131	955	912025	30·903	97·724
906	820836	30·100	95·184	956	913936	30·919	97·775
907	822649	30·116	95·237	957	915849	30·935	97·826
908	824464	30·133	95·289	958	917764	30·952	97·877
909	826281	30·150	95·341	959	919681	30·968	97·929
910	828100	30·166	95·394	960	921600	30·984	97·980
911	829921	30·183	95·446	961	923521	31·000	98·031
912	831744	30·199	95·499	962	925444	31·016	98·082
913	833569	30·216	95·551	963	927369	31·032	98·133
914	835396	30·232	95·603	964	929296	31·048	98·184
915	837225	30·249	95·656	965	931225	31·064	98·234
916	839056	30·265	95·708	966	933156	31·081	98·285
917	840889	30·282	95·760	967	935089	31·097	98·336
918	842724	30·299	95·812	968	937024	31·113	98·387
919	844561	30·315	95·864	969	938961	31·129	98·438
920	846400	30·332	95·917	970	940900	31·145	98·489
921	848241	30·348	95·969	971	942841	31·161	98·539
922	850084	30·364	96·021	972	944784	31·177	98·590
923	851929	30·381	96·073	973	946729	31·193	98·641
924	853776	30·397	96·125	974	948676	31·209	98·691
925	855625	30·414	96·177	975	950625	31·225	98·742
926	857476	30·430	96·229	976	952576	31·241	98·793
927	859329	30·447	96·281	977	954529	31·257	98·843
928	861184	30·463	96·333	978	956484	31·273	98·894
929	863041	30·480	96·385	979	958441	31·289	98·944
930	864900	30·496	96·437	980	960400	31·305	98·995
931	866761	30·512	96·488	981	962361	31·321	99·045
932	868624	30·529	96·540	982	964324	31·337	99·096
933	870489	30·545	96·592	983	966289	31·353	99·146
934	872356	30·561	96·644	984	968256	31·369	99·197
935	874225	30·578	96·695	985	970225	31·385	99·247
936	876096	30·594	96·747	986	972196	31·401	99·298
937	877969	30·610	96·799	987	974169	31·417	99·348
938	879844	30·627	96·850	988	976144	31·432	99·398
939	881721	30·643	96·902	989	978121	31·448	99·448
940	883600	30·659	96·954	990	980100	31·464	99·499
941	885481	30·676	97·005	991	982081	31·480	99·549
942	887364	30·692	97·057	992	984064	31·496	99·599
943	889249	30·708	97·108	993	986049	31·512	99·649
944	891136	30·725	97·160	994	988036	31·528	99·700
945	893025	30·741	97·211	995	990025	31·544	99·750
946	894916	30·757	97·262	996	992016	31·559	99·800
947	896809	30·773	97·314	997	994009	31·575	99·850
948	898704	30·790	97·365	998	996004	31·591	99·900
949	900601	30·806	97·417	999	998001	31·607	99·950
950	902500	30·822	97·468	1000	1000000	31·623	100·000

INDEX

THE MECHANICAL AND PHYSICAL PROPERTIES OF THE BRITISH STANDARD EN STEELS

(B.S. 970—1955)